NATURE'S GREENHOUSE:

OMINOUS DEVELOPMENTS

ALEX COOK and J. GEORGE

Outskirts Press, Inc.
http://www.outskirtspress.com

ISBN: 978-1-4787-4205-0

Outskirts Press and the "OP" logo are trademarks belonging to Outskirts Press, Inc.

PRINTED IN THE UNITED STATES OF AMERICA

DENVER, COLORADO

Author's Note

The observed accelerated melting of glaciers and polar ice requires an increased absorption of immense amounts of heat energy on the Earth's surface. This phase change from ice to liquid occurs at a constant freezing point temperature. The independent observation of a global average surface temperature increase in the last 100 years is about 0.8°C (1.44°F). There is a resultant sea level rise of a bit more than an inch per decade from the meltwater addition and the thermal expansion of the oceans. The related measurements of atmospheric CO_2 greenhouse gas have shown concentrations of 400 parts per million; this increase of 43% since the industrial revolution is remarkably consistent with the burning of fossil fuels. This is the anthropogenic atmospheric greenhouse effect in action.

Recent extreme weather events are a more subjective observation. The label of climate change for Superstorm Sandy, the Midwest and California droughts, or the Colorado wildfires and floods is suggestive but debatable. These uncertainties are used by the fossil fuel industries in their well-funded propaganda to sow doubt about climate change in the public. Our television news media follow the advertising money in ignoring the subject, leading a significant fraction of the public to continue to disbelieve climate change. Our education system fails to exercise leadership and avoids mention of the topic. Government of the people fails to act, since such action would imply that the electorate is aware of a need for action. Some of our government leaders claim climate change is a fraud. Others cite economic difficulties as the excuse for procrastination in conversion to renewable energy: we must not jeopardize jobs or increase the cost of electricity. But the climate adaptation costs of beach erosion, water desalination, firefighting, post-tornado reconstruction, and rising supermarket food prices are alternate hardships we all face as we work our way to that arbitrary tipping point of a global 2°C temperature rise.

In this presentation of the environmental influence of our increasing human population, we have focused on the story of the science of global warming; this is the driver of Earth's response to further climate changes. There can be no debate of the validity of the 100 year established radiation science, nor the warming of the Earth's surface by the greenhouse trapping of heat by its atmosphere. The scientific validity of the increased atmospheric CO_2 greenhouse gas from the burning of fossil fuels is firmly established. We believe the reader's familiarity with the logic of these fundamentals will inspire an objective analysis of the ongoing climate change. Then the human response to this most dangerous environmental threat may become more constructive and efficient.

Preface

There are more than seven billion humans on planet Earth. Only the insects outnumber us. Humans have the ability to adapt to the geographic and variable extremes of temperature that the planet has to offer. Nature is not an independent companion. Our activities in moving about, collecting food and using our natural resources make us an integral part of the planet's ecosystem. We have learned to harness available supplies of cheap energy, often without regard to environmental damage. Recognition of some of these environmental mistakes has generated some governmental regulation, but we are apparently loath to follow the fundamental science in correcting that most dangerous environmental threat, global warming and climate change. Still, we are uniquely qualified with the intelligence to imagine our future, and we have begun to realize our limits of sustainability. Do we have the moral integrity to modify our mindless pollution and avoid future deleterious consequences?

When the human population was only a fraction of that at present, we could live without regard for our environment, short of all-out nuclear war. And we could simply accept and enjoy our unique climate with little understanding or care. But no longer. It is time to wake up and understand and appreciate the controlling laws of nature. To say it's too hard or I don't want to think about it is an arrogant, lazy, and stupid attitude. We hope that in this composition we might be successful in helping those who feel the beginnings of responsibility to understand the wonderful life-giving planet we have and how to care for it. Recent episodes of extreme weather should have inspired some attention; however, those so unfortunate as to be directly involved will be completely occupied with getting their lives back together. There are other individuals who must devote their entire lifetimes to finding food and shelter; they will have no opportunity to give thought to such abstract things. And there are those more fortunate who are content to be fed and entertained and who will experience no concern on learning of the climate threat. They will unknowingly comprise the bulk of the human lemmings scrambling over the cliff into the rising seas if we fail to convince our leaders that they must address this imminent danger. But you have to help; you may find yourself on a slippery slope into science with words like photosynthesis—and energy!

The role of science in human civilization has always been outside our traditional tribal organization. A few hundred years ago a tribe used their Earth-centered religion to hold its members together. The introduction of modern science by Galileo didn't fit. The various human constructs of religions, economies, and governments have since been somewhat modified and continue to change on a random and slow schedule. But the laws of science are not amenable to such whimsical behavior. Their constraints must always be a factor in our responsibility to care for the planet. Those citizens who completely disregard the scientific facts to follow the misguided policies of their religion or political party threaten to disrupt the Earth's stable climate and the habitat of future generations.

We hope that our story of the science of global warming and climate change might serve as a basis for the intellectual background of our nation's voters. We may then expect that our government of, by, and for the people may understand and respond positively to the demands of its electorate. Many citizens have the basic smarts to be successful in business or politics, and there are a few who have the intelligence to lead in times of stress or danger. Perhaps we might succeed in our effort so that an accurate knowledge of the fundamental science of the greenhouse effect and its control of climate change will aid those in leadership positions in education, media, government, and religion to initiate corrections in the care for our environment.

The 2007 report of the United Nations Intergovernmental Panel on Climate Change (IPCC) affirmed a 0.74°C global temperature increase in the 100 years ending in 2005.

The release of the United Nations Intergovernmental Panel on Climate Change (IPCC) 2013 "Summary for Policymakers" has strengthened the conclusions of the 2007 report. The evidence for human influence on climate change has grown. It is **extremely likely** that human influence has been the dominant cause of the observed warming since the mid-20th century.

The World Meteorological Organization (WMO) reports that the monthly concentrations of carbon dioxide (CO_2) in the atmosphere topped 400 parts per million (ppm) in April 2014 throughout the northern hemisphere.

Publication of the 2014 "Climate Change Evidence and Causes" by the US National Academy of Sciences and UK Royal Society confirms these conclusions, and the 2014 U.S. National Climate Assessment "Climate Change Impacts in the United States" presents a complete and authoritative story of this environmental situation in our country.

Table of Contents

1 | Earth's Radiation Budget

Planet Earth has enjoyed a stable climate for thousands of years; paleontologists called it the Holocene. A large variety of life forms, including humans, has evolved and thrived in these conditions. *Homo sapiens*, in particular, has evolved a superior intelligence that has enabled the species to develop a variety of civilizations that clearly dominate all other creatures. The natural world of plants, animals, sea life, and mineral resources has been the reservoir exploited by an increasing population of complex human societies. Earth's climate is unique in the solar system and perhaps in the universe. A fundamental characteristic of this climate is its temperature.

You are wondering why you should give any attention to this old retired science codger. What is his authority to say he has all the answers? There are lots of other folks with opinions galore, even other scientists. That's good—question authority! And there are some nonscientist journalists who have carefully studied the scientific facts, like Bill McKibben. Al Gore is the best known of these journalists, but many people confuse his journalistic efforts with his political persona and are unlikely to give his books, the Nobel prize-winning *An Inconvenient Truth*, and *Our Choice: A Plan to Solve the Climate Crisis*, and articles in The New York Times, Rolling Stone, and elsewhere the attention they merit. And there are lots of individuals with a science background—like medical doctors for example. I wonder, would you ask an atmospheric scientist from NCAR for advice about your fever and cough? Fortunately, the reply would likely be, "Go see your doctor!" But I doubt that the doctor would say, "Read the works of Jim Hansen". The real experts could give you much more detailed information, but my efforts should be sufficient to activate your little gray cells.

I have been inspired by liberal libations of Burnett's Gin to assume the vita of that not very famous atmospheric scientist of identical name who has published 33 years of atmospheric measurements related to stratospheric ozone in refereed journals. But more importantly, I can read and understand the published research on global warming and climate change. I am now using the pen name Alex Cook.

I'm sure you are aware that you will not learn much from our established news organizations who generally censor their news to eliminate the words "climate change" and "global warming" as profanities that would cause the loss of substantial support from the fossil fuel industry. If forced to discuss the possibilities or consequences of climate change, they insist on the policy of airing both sides of this issue with equal time, regardless of the fact that polls of atmospheric scientists show that 97% of them agree that anthropogenic climate change is happening. And never do we hear details of the fundamentals of atmospheric science; the hand-waving result of computer mathematics is deemed sufficient. I know it is difficult to ignore the expensive misinformation from the fossil fuel deniers and their practice of calling climate change and global warming a fraud or a hoax. I don't have a well-funded publicity organization, but I get a monthly Social Security check that is indicative of a certain amount of wisdom from a lifetime of experience in the science profession. So! Trust me, dammit! We're going to present the story of the atmospheric greenhouse effect that guarantees the validity of global warming and climate change!

Actually, this information has been around for over a hundred years, but it's not common knowledge. The genie that we call the

Greenhouse Effect has had a benign influence on the development of humanity, but now our increasing numbers have let him escape from the bottle. This little fellow has the power to make our life very inconvenient with disruptive changes in our climate. So, wouldn't you like to be able to convince your kids that you're capable of helping them with their homework if their teachers should actually mention global warming and climate change, and that you are giving attention to your offspring's world after they leave the nest?

And you, my dear reader, have heard this all before and are about to switch on the TV and surf the channels. You don't want to suffer an exposure to science from some ancient professor. Never mind, he's grown weary with writing scientific manuscripts that no one reads. So Alex Cook and J. George have made the commitment to present some words of wisdom about global warming and climate change.

OK. Alex Cook is in charge here. Let's have a bit of a quiz to see if you're prepared to handle this stuff. If the neon sign in front of the downtown bank reads 25°C, what is the temperature in °F? No, you don't have time to look it up on the Internet. And Al Gore's UN friends say the global temperature has increased 0.74°C; what does that really mean?

Yes, the temperature at the bank is 77°F, but I bet you cheated by looking at the thermometer outside the kitchen window! Most of the country is stuck on the Fahrenheit scale in which water freezes at 32° and boils at 212°—at sea level, that is. Most scientists use the Celsius scale with those temperatures reading 0° and 100°; isn't that neater? And between those fixed calibration points there are 180 divisions on the Fahrenheit scale with 100 divisions on the Celsius. So 0.74°C is 1.33°F. (Fahrenheit divisions are about 10% less than doubled those of the Celsius scale.) Our average measured global temperature is 15°C, or 59°F.

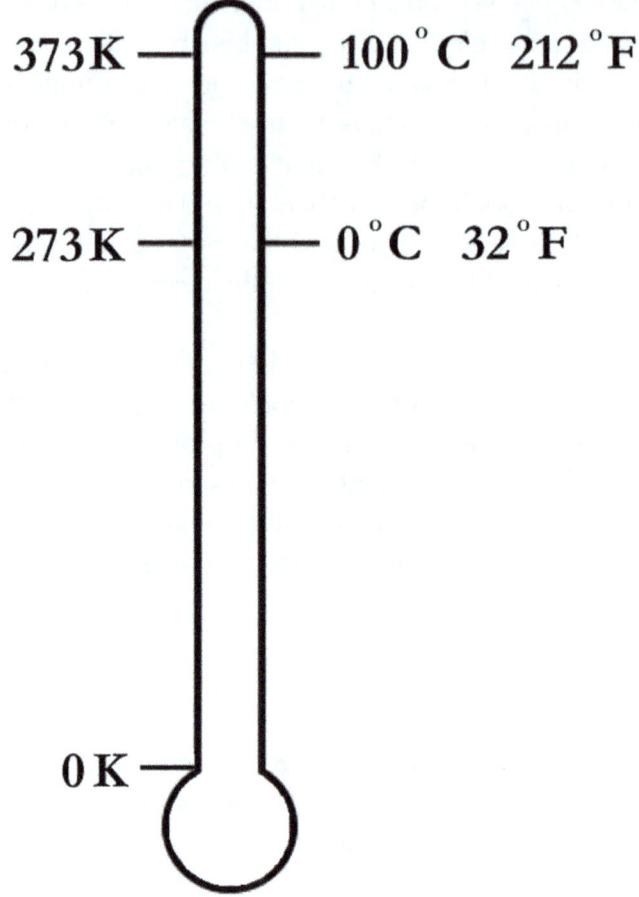

Fig 1. Temperature scales used to measure the energy of random molecular motion (heat energy).

But those are just numbers. What do they really mean? Science books say that temperature is a measure of the random energy of motion (heat energy) of the surroundings. And the number tells us which direction the heat will flow; that is, whether we will feel warm or cold. The science gets even neater if we use the Kelvin scale, which measures the total heat energy; it reads 0 when there is no random motion. The scale divisions are the same as Celsius, and water freezes at 273 K. (Now there will be some nitpicker out there who says 273.15 and that there is still some quantum energy left over when the heat is gone, but that won't affect our discussion about the Earth's climate.)

OK, we're on the same page with temperature scales, and our air conditioners work just fine with today's average global temperature. But our island neighbors complain that we've been warming things too much with our gas guzzling cars and jumbo jets, and it's beginning to make life a bit inconvenient with rising sea levels. And there are episodes of extreme weather all over the globe that make us wonder if the climate is changing. Every aspect of life on this planet, of plants, animals, and humans, is vulnerable to this increasing temperature. But isn't Mother Nature in control of that temperature? Actually, yes. The laws of natural science determine our global temperature. The Earth has an energy account that is in a continual state of flux. There is a continuous deposit of energy from the sun, and the Earth works to establish a temperature that will balance that radiation. Our ancestors learned the precise science that describes this process, but currently there is a slight perturbation in a few of the parameters because so many humans have gotten into the act. Don't you think we should understand the rules of this game if we insist on participating?

Energy can be transported by radiation; we call it heat radiation. It originates in basically the same manner as your TV signal—by vibrating electric charges, just in a different part of the spectrum of wavelengths. Our ancestors made accurate measurements of the heat wavelength spectrum over a hundred years ago—measurements before a theory was established. And the theory that agrees, proposed by Max Planck, was really a quantum step forward, the original introduction of the quantum theory of matter, now the basis for some 35% of our toys, from cell phones to photovoltaic solar panels to supercomputers. The fundamental idea of discrete packets (photons) of energy proportional to the frequency of radiation ($E=hf$) ($f=c/\lambda$) is a fundamental link in the explanation for the temperature dependence of the continuous heat spectrum.

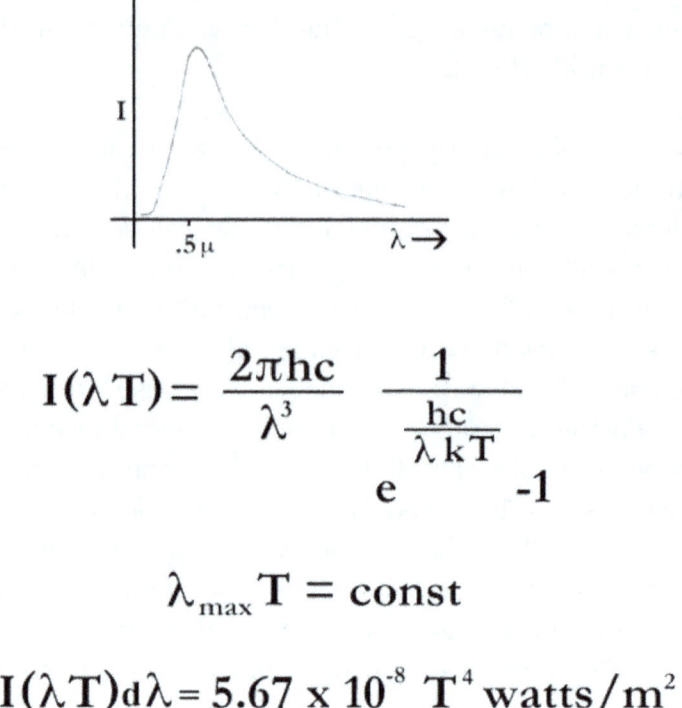

$$I(\lambda T) = \frac{2\pi hc}{\lambda^3} \; \frac{1}{e^{\frac{hc}{\lambda kT}} - 1}$$

$$\lambda_{max} T = const$$

$$\int I(\lambda T) d\lambda = 5.67 \times 10^{-8} \, T^4 \, watts/m^2$$

Fig 2. The intensity spectrum of heat energy.

The intensity rises abruptly at short wavelengths to a maximum, with a gradually decreasing tail at long wavelengths. This spectral measurement is the bottom line in the science. However, the theoretical Planck's spectral distribution $I(\lambda, T)$ is useful for temperature calculations since it is completely dependent on the temperature of the source (the temperature is on the Kelvin scale).

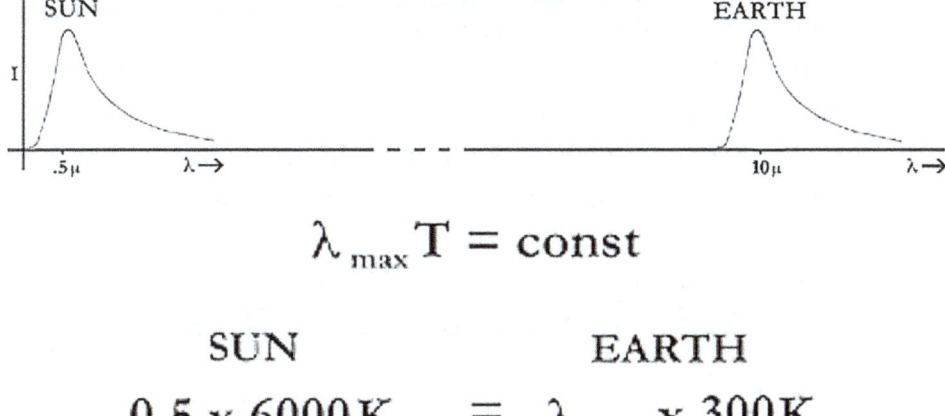

$$\lambda_{max} T = \text{const}$$

$$\text{SUN} \qquad\qquad \text{EARTH}$$

$$0.5 \times 6000\,\text{K} \quad = \quad \lambda_{max} \times 300\,\text{K}$$

$$\lambda_{max} = 10\,\mu$$

Fig 3. The intensity shift of Earth's heat radiation calculated from the general mathematical formula of heat radiation.

If we do a mathematic examination of Planck's formula for the radiation, we find the wavelength for the maximum intensity times the Kelvin temperature is constant for any source temperature. That is:

$$\lambda_{max} T = \text{const}$$

And the total intensity of heat radiation, calculated from Planck's formula, is given by

$$5.67 \times 10^{-8} \cdot T^4$$

We can use that expression for the maximum intensity of the heat radiation and our observations of the sun to predict the wavelength of maximum heat radiation from the Earth. I will use approximate values to simplify the arithmetic (good nuff for govmnt work). We have maximum intensity for the sun at 0.5 microns in the visible and sun surface temperature of 6000 Kelvin to find the wavelength of maximum emission from Earth with a temperature of 300 Kelvin:

$$\underset{\text{Sun}}{0.5 \text{ microns} \times 6000\text{K}} = \underset{\text{Earth}}{\lambda_{max} \times 300\text{K}}$$

We find the maximum Earth radiation at a wavelength of about 10 microns in the infrared.

Radiation Balance

The energy balance of the Earth is determined entirely by radiation; the planet has no other means of gaining or losing heat energy. The spectral distribution follows the general rules described above with the wavelength of the maximum intensity controlled entirely by the temperature of the source. The radiation from the hot surface of the sun has its maximum in the visible. We detect it every day with our eyes, and we can feel its absorption on our skin. The heat from the red coals of our wood fire is also visible, but the radiation from the slightly warm Earth surface is beyond the sensitivity of our eyes in the long wavelength infrared.

If you haven't become bored with my tedious discussion of the necessary details, we can get down to the nitty-gritty of a calculation of the Earth's radiation budget. We will set the measured solar intensity absorbed on the Earth equal to the temperature-dependent formula of the Earth's outgoing radiation. You can do the calculation for the average temperature of the Earth's surface with paper and pencil if you remember how to do arithmetic. We use the satellite measurement of solar intensity at normal incidence at the top of the Earth's atmosphere. This disc of the circular Earth target perpendicular to the radiation has area of πr^2, but the 24-hour average of radiation on the Earth is distributed over a spherical area of $4\pi r^2$. So the satellite measurement corrected by a factor of 4 for the spherical surface of the Earth gives the average intensity on the Earth to be

$$1370/4 = 343 \text{ watts/m}^2 \quad (\text{m = meter})$$

The mathematical expression for the total intensity of heat radiation from the Earth, as found in any physics textbook, is given by

$$(5.67 \times 10^{-8} \text{ watts/m}^2\text{K}^4) \times T^4$$

We will use this to calculate the Earth's radiation temperature, T. Actual measurements show that about 30% of the solar energy, called the albedo, is simply reflected back into space by the Earth's surface, atmosphere, and clouds. If we make the simplified assumption that the atmosphere has no other effect on the radiation balance, we can make a preliminary calculation of the energy balance. Using 0.7×343 watts/m^2 as the absorbed intensity, I calculate T = 255 K for the radiation surface of our planet.

This result for our calculated average Earth radiation temperature on the Celsius scale is -18°C, slightly below 0°F. But the average measured temperature at the surface of the Earth is 15°C =288 K (59°F). That is a difference of 33°C. Not even close! There are more recent measurements with slight changes for the input quantities, but the conclusion is essentially unchanged.

The calculation is too simple to be a careless mistake in arithmetic. How important is it? One might argue that on the Kelvin scale the calculated radiation temperature is only slightly less than 10% below the measured average total heat energy of the Earth's surface. Is that important for Earth's plants and creatures? Only if they can live in a world of ice! So we have a wondrous mystery of how planet Earth has developed and maintained life at a global average temperature of 15°C at its bosom while it must radiate energy into space at –18°C.

(Ah, but perhaps we are comparing apples and oranges. If we were to imagine that the effective radiation surface is high in the cold atmosphere instead of at the Earth's surface, there would be no discrepancy...but that explanation must await our discussion of the atmospheric greenhouse effect.)

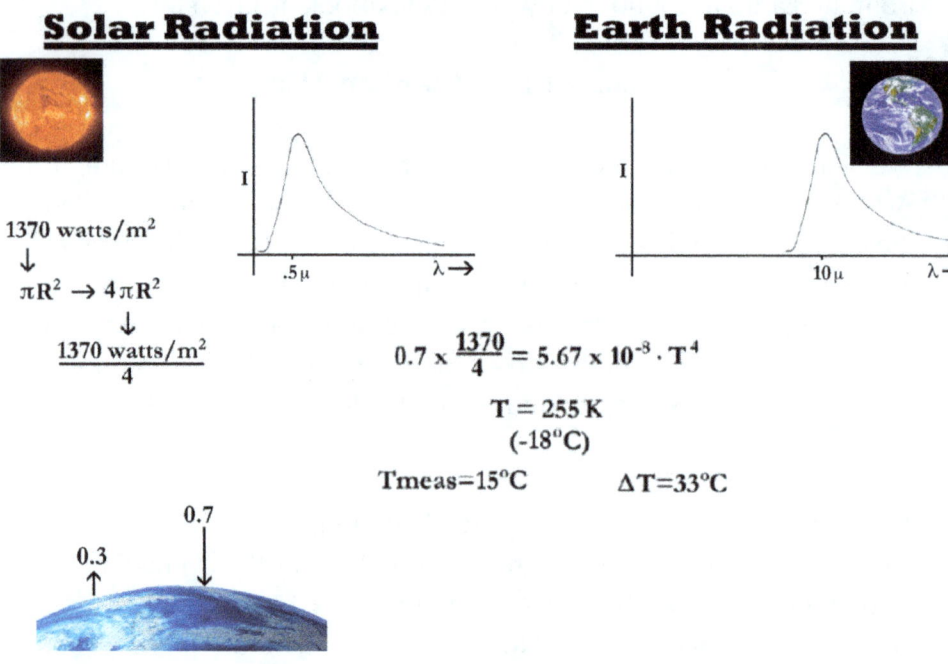

Fig 4. Calculation of Earth's radiation balance from satellite measurement of solar intensity and the general mathematical formula of heat radiation.

2 | The Evolution of Life

I am very much aware that at a certain temperature ice melts, and at an increasing temperature the water molecules move apart in expansion, and escape from the surface by evaporation. However, in avoiding more formal instruction beyond covalent and hydrogen bonds, I am perhaps somewhat life science-challenged. Therefore, I have sought help from my companion who has the knowledge and professional experience in such matters. She answers readily to "George" when behind the wheel of my Jeep, in honor of that very efficient chauffeur of the fictitious Mrs. Bradley, introduced by the author Gladys Mitchell. Please give her your careful attention so that you may understand how we and our fellow creatures must respond to climate change in this new epoch that is now called the Anthropocene.

Hello. My name is George....Just George. The professor has asked me to add my thoughts to his opus. Our backgrounds are complementary—his in physics and atmospheric science and mine in microbiology and botany—and he feels that together we have a broader and more comprehensive point of view.

Planet Earth appears to us humans to be a perfect place to live, and for us, it is. It has the right surface temperature, the right balance of atmospheric gases, the right gravitational pull, and the right resources to nourish all the inhabitants necessary to maintain the system. From its cosmic beginnings, billions of independent alternatives came together in a specific way to produce the conditions we enjoy. It is no coincidence that these conditions are perfect for us because we evolved in response to these parameters. We are uniquely suited for life on our unique planet. We cannot migrate to other planets, and, conversely, it is unlikely that life forms from other planets have evolved to fit our conditions on Earth. If there is life on other planets, as many scientists and all science fiction writers believe, it has evolved to fit the parameters of its home planet and would likely be unrecognizable to us.

Planet Earth is unique and remarkable, as are we who have evolved here. We have adapted to develop advanced civilizations in its various temperature regions and have managed to live in the extremes of tropic and polar regions. We have developed a moral code and have learned to live cooperatively and peacefully, although not always without conflict and disagreement. We have learned to cultivate crops in favorable places and have developed nitrogen-based fertilizers to increase yields and support growing populations. We have learned to harvest other creatures as a food source on land, in the air, and in the sea. We have learned to exploit mineral resources and fossil fuels. Our technology has eased the effort and time necessary for subsistence, and we have the leisure time to develop the arts and other pleasurable pastimes.

However, our numbers and our technologies have allowed us to turn the corner, so to speak, and we humans are now in a position to drive what happens next. We are no longer responding to what nature has provided for us; instead, we are driving the responses of nature, most notably changes in our climate resulting from our persistent use of fossil fuels that pollute our atmosphere with an increase in heat trapping gases. Perhaps this is the next step in planetary evolution: we humans, who too often arrogantly assume we are the end product of evolution, will either have to adapt to new conditions or rectify our abuses of the present environment. We should feel a responsibility to our descendents that should prompt us to make carefully studied decisions. Our stubborn determined shortsightedness is leading us down a dangerous path to a future incompatible with our present society. The appropriate response to climate change might cost a lot, but refusing to act now will cost us the Earth.

Back to Alex. He has assembled the story of the real physical science that goes into those mysterious computerized climate models.

3

The Greenhouse Effect

If you have done any serious reading on global warming, you have seen the results of that simple calculation of radiation balance presented as a statement of fact; I've just shown the details to you. And the result will always be valid so long as our sun's nuclear fusion power continues at the present rate and Jupiter's perturbation of the Earth's orbit does not reoccur until the far distant future.

My purpose in explaining radiation balance was to demonstrate mathematically that the Earth does not suffer heat radiation loss directly to space from the surface. Clearly, the atmosphere has some other active mechanism that traps much of the surface heat energy. This is not a new discovery by NASA and myself; the science was reasonably well understood a hundred years ago. But most of the planet's inhabitants simply accept the situation as an invariable fact of Nature. Now that we have gone forth and multiplied, we have become an active and influential player in Nature's game. So it's time to learn the rules!

Scientists have known for a hundred years that certain trace atmospheric gases like water vapor and carbon dioxide are transparent to the incoming visible radiation but are strong absorbers of some of the outgoing heat radiation in the infrared. The situation is similar to the functioning of greenhouses constructed to grow vegetables and flowers. These buildings usually have glass roofs that permit transmission of the visible sunlight but absorb and trap the infrared heat inside the structure. (The fact that the walls do not let the warm air escape is at least, if not more, important. And it may also be necessary to employ fans or windows to control any excess temperature.) But the Earth communicates with the vacuum of space only by radiation. The Earth atmosphere transmits most of the incoming 70% of the solar radiation that was not reflected. (The solar ultraviolet is absorbed by oxygen and ozone, protecting us from sunburn by these high-energy photons, and some solar infrared is absorbed by water vapor and other molecules. These are small fractions of the total sunlight, however; maximum intensity is in the visible.) Most of the outgoing radiation from the cooler Earth is in the infrared, where it is strongly absorbed at certain wavelengths by water vapor (H_2O), carbon dioxide (CO_2), methane (CH_4), nitrous oxide (N_2O), ozone (O_3), CFCs, and other molecules with complicated structures. These molecules have low-lying closely-spaced energy levels determined by the vibration and rotation motions of their atomic constituents. Their electronic structure permits each of these molecules to behave like a very efficient dipole antenna for absorption and emission of radiation. The low energy infrared photons from the Earth are absorbed very neatly into these energy slots. This "atmospheric greenhouse effect" of transmission of incident visible solar radiation and infrared absorption of Earth heat radiation by trace constituents of the atmosphere is the missing factor responsible for trapping the additional heat that gives the Earth's surface a temperature of 33°C above the temperature of an effective radiation surface that must be high in the cold levels of the lowest layer of the atmosphere (the troposphere).

Accurate laboratory measurements give us the probabilities of narrow absorptions in the infrared spectral wavelength region by water vapor, carbon dioxide, methane, nitrous oxide, and other molecules of trace concentrations. Water vapor absorbs strongly in narrow regions starting at a wavelength of about 1 micron, becoming continuous at about 15 microns. (A micron is one millionth of a meter.) Carbon dioxide absorbs at discrete wavelengths in the 2 to 8 micron region and has a very strong absorption at 15 microns. None of these molecules absorb strongly in the 10-12 micron region; we call it an atmospheric window.

There is also a strong ozone absorption all by itself at about 10 microns. (But don't get confused like many of your neighbors. This infrared greenhouse property of the ozone molecule is independent of its important role in the absorption of skin-damaging ultraviolet. It is the same molecule; it is found in both the high and low atmosphere (stratosphere and troposphere), and where we find it is of different importance. Keep your shirt on; I'll make you an expert later.)

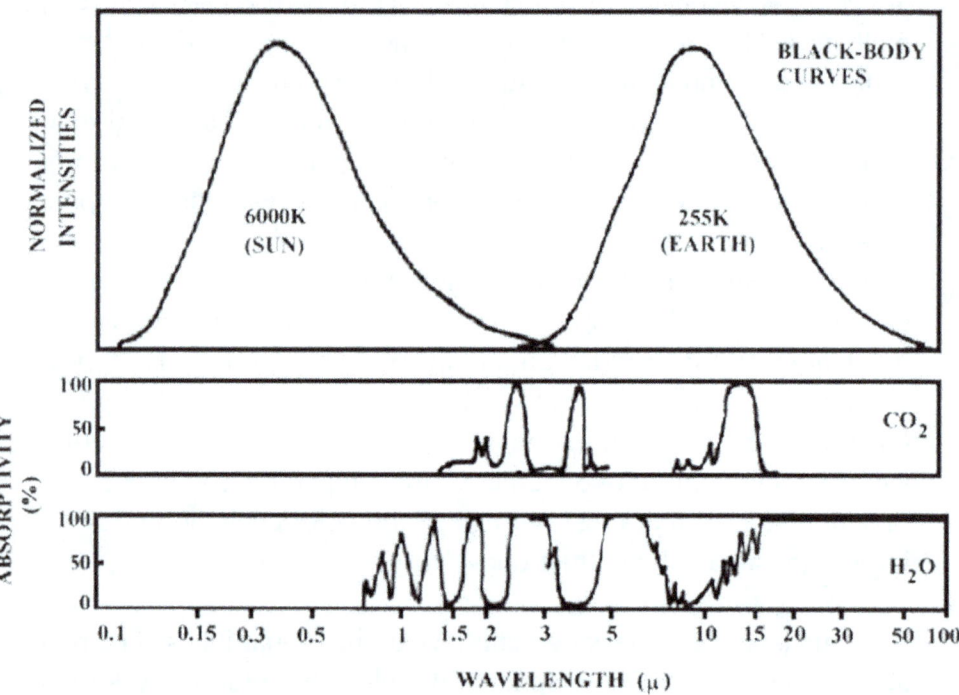

Fig 5. Measurements of the infrared absorption spectra of the greenhouse gases. (Note that the wavelength scale is nonlinear (logarithmic-factors of 10 now increase linearly). Units are microns (millionths of meters).

These greenhouse gas absorptions occur just where the Earth's infrared radiation maximizes for surface temperatures. In the lower atmosphere, some of the absorbed energy is shared with the atmosphere in molecular collisions; most is reradiated from the greenhouse gas in all directions. The downward fraction furnishes the warming radiation blanket that did not exist in the absence of an atmosphere that we considered in our earlier calculation.

Now you say, "So we've got the atmospheric greenhouse effect that gave us the Holocene climate. And because most of the Earth is covered with oceans, this should be dominated by water vapor, the fourth most abundant atmospheric constituent. So I'll be a denier and insist that humans can't be affecting the climate with the small concentrations of those other gases. We're not changing Nature."

You are correct up to a point. The greenhouse effect of all that water vapor is several times that of the other trace constituents. And I must confess to the initial naive assumption that a few other contributions wouldn't matter much. Muddy thinking!

Feedback

Ah, but atmospheric science is not as simple as we thought. Those trace constituents have always played a significant role in determining the global temperature. You see, the oceans covering 70% of the Earth's surface are warmed by the slight additional radiation trapped by the trace constituents mentioned above. This warms the atmosphere and the oceans a bit more and causes a slight additional increase in evaporation of water from the oceans. The water vapor greenhouse effect gets a boost, the atmosphere is warmed some more, and we do it all again. A few more cycles of this process moves the whole system to a higher temperature. We call this the water vapor feedback. Professor Alley from Penn State has called the CO_2 step in this process the Control Knob on the thermostat for that additional kick in the greenhouse effect by increased water vapor.

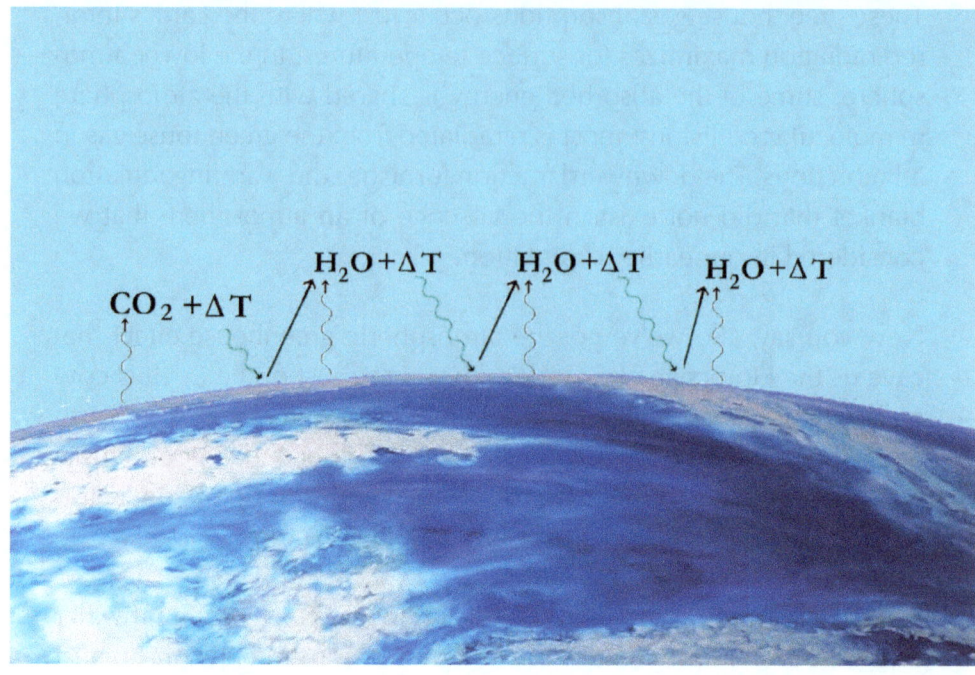

Fig 6. The CO_2 control knob on the thermostat for water vapor feedback in the atmospheric greenhouse effect.

Perhaps it is useful to compare this CO_2–H_2O feedback with compound interest on your bank savings account. It may be just a few percent interest—it doesn't sound like much. But in about 20 years the principal will have doubled! Gotcha!

So now we have very good reason to expect that the final greenhouse controlled temperature of the Earth has always been greatly increased to our present comfort zone by the evaporation of extra water vapor in response to the additional trapping of heat by trace molecules like CO_2. We finally have an explanation for that large atmospheric contribution of 33°C trapped surface heat that gives us the present climate suitable for our lives and the rest of God's flora and fauna.

The Effective Radiation Surface

There are certain other facts of the atmospheric structure that I'm sure you are familiar with—but I'll remind you anyway. Our atmosphere is composed mostly of molecules of nitrogen and oxygen. Each of these molecules is composed of a pair of identical atoms; this does not make an electric dipole and they do not absorb infrared radiation. Since these atmospheric gases are compressible, the weight of the atmosphere yields a maximum molecular density at the Earth's surface, decreasing exponentially upward. Ground level heating by solar radiation produces convective lifting that moves parcels of air to elevated regions of lower pressure; expansion uses some of the gases' thermal energy, and the temperature lapse rate of dry air shows a decrease of 10°C/kilometer, 6°C/km for moist air (average observed about 6.5°C/km or 3.6°F/1000ft). The principal greenhouse gas, water vapor (H_2O), is mixed throughout the lower atmosphere (the troposphere). It condenses to form clouds in the cool air, giving us our weather, and it is trapped as ice crystals in the cold air at the top of the troposphere. The air in the region above the troposphere, the stratosphere, is dry. On the other hand, the trace amounts of other greenhouse gases like CO_2 are not trapped, and are mixed throughout the entire atmosphere.

That part of the Earth's radiation that was trapped by the greenhouse molecules like CO_2 and H_2O and radiated back to the Earth's surface is an additional atmospheric effect that was neglected earlier. This warm radiating blanket gives us the warm surface temperature averaging about 15°C that plants and animals—including humans—have evolved to utilize. And as the Earth's radiation continues upward by the greenhouse gas absorption and re-radiation, at about 15-20,000 ft, where the molecular density is low, the radiation is no longer efficiently trapped by the greenhouse molecules and may escape to space. And independently, the lapse rate discussed above has established a temperature at this altitude near that –18°C that we calculated earlier. This is the Earth's fuzzy effective radiating surface that yields the Earth's energy balance.

Now you are no doubt aware that CO_2 is cycled in and out of the atmosphere by the trees in photosynthesis and decay. So in the recent past we had a rather marvelous set of activities with just the right amount of atmospheric CO_2 to control the H_2O that gives the trapped heat for a surface temperature of 15°C. And this atmosphere has just the right lapse rate for the effective radiation surface at −18°C to give the balance of incident energy from the sun with outgoing Earth radiation. This background CO_2 concentration observed prior to the industrial age and this natural science thermal property of our gaseous atmosphere was critical for the stability of Earth's greenhouse effect in warming the surface to 15°C, yet balancing the radiation at a surface of −18°C.

There have been perturbations of this climate situation by volcanoes, solar activity, and earth orbit changes. And we could devise a theory for the planet's successful response to these energy imbalances, something short of fairies in the environment, but we do not have the measurements of forest or plankton growth and decay to test such a theory. But now we have a human caused perturbation of rapid increase of greenhouse gas emissions that surpasses anything in our recent past. Our only realistic hope for a correction is a human understanding and solution.

The strong water vapor absorption in the wavelength region near 6 microns is not trapped by the low-density water vapor and this energy is lost to space. The water vapor satellite pictures shown on TV are the result. And there is the atmospheric window through the infrared

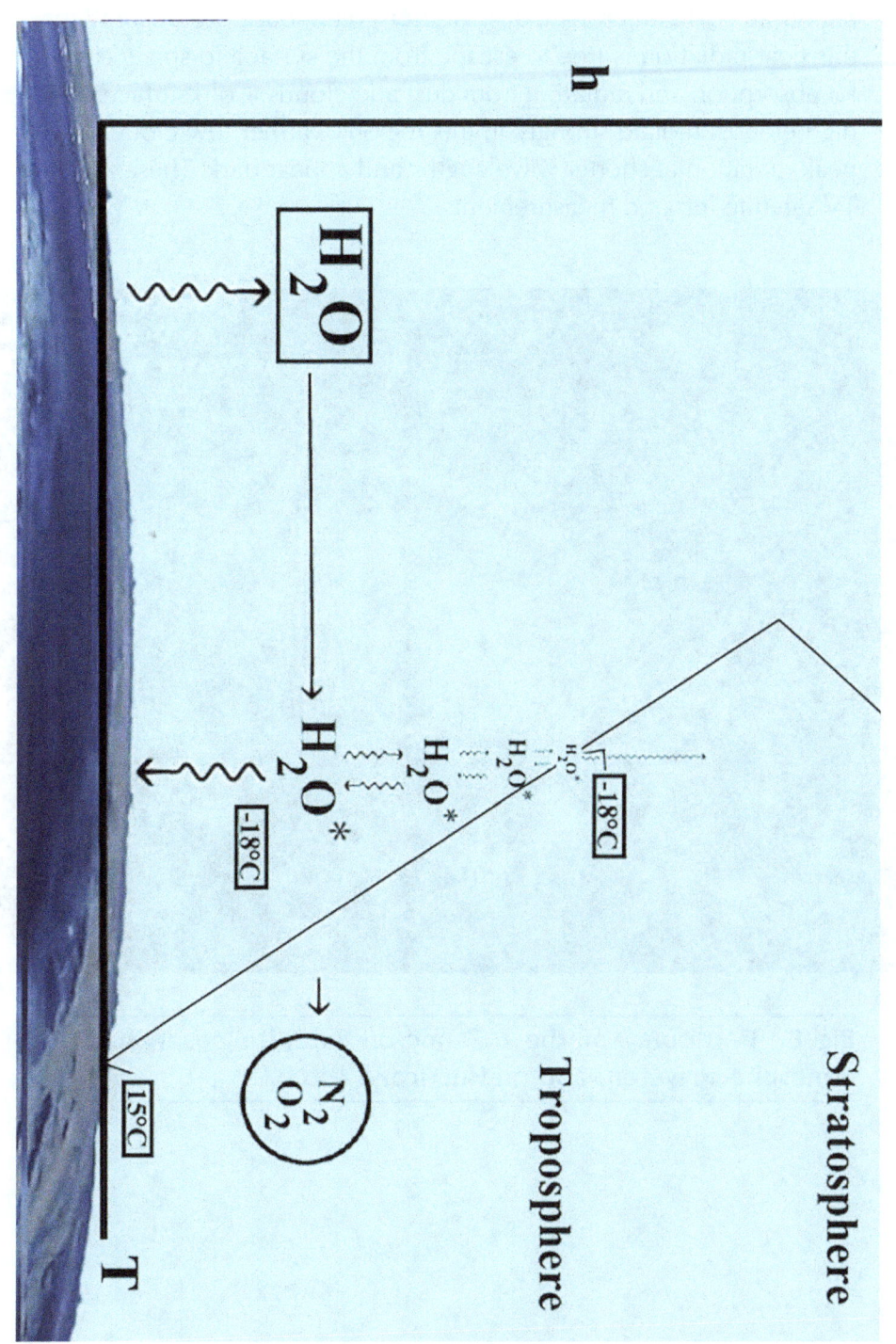

region at 10-12 microns that is free of greenhouse gas absorptions; this heat radiation is free to escape from the surface to space except for absorption and radiation from dust and clouds. Cold surfaces, like high clouds, radiate strongly in this region; warmer low clouds have peak radiation at shorter wavelengths and appear dark. These are the TV satellite infrared measurements.

Fig 8. TV display of the 6.7 micron satellite observation of atmospheric water vapor in Hurricane Mitch.

Fig 9. TV display of the 12 micron infrared satellite cloud observation; cold high clouds are bright, warm low clouds are dark.

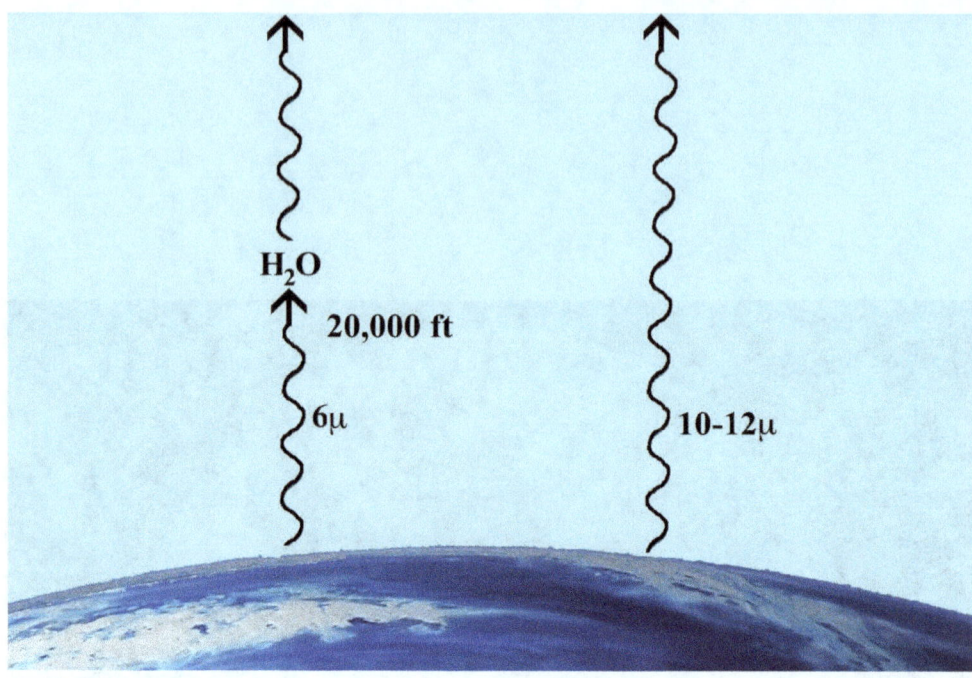

Fig 10. The radiation processes for the TV displays of satellite observations of Earth's radiation loss to space.

With this situation of low density in the upper troposphere, the CO_2 greenhouse gas also reradiates its energy in all directions to space in a manner similar to that described for the water molecules. And because these molecules are not trapped by the cold temperatures, they continue to diffuse above the troposphere and to lose radiant energy in the stratosphere and above; in fact, the upper atmosphere is actually cooled below that expected for its upper atmospheric chemistry. So, you see, the Earth's effective radiation surface is necessarily somewhat fuzzy, and these details vary over the Earth's surface.

This is greatly oversimplified, of course, in that there are many other factors that complicate the energy balance. There is reflection, absorption, and radiation from clouds, dust, and aerosols. The surface albedo varies as the snow melts seasonally, and with global warming. Various

	T Surface	T Radiation balance	ΔT Greenhouse Effect
Mars 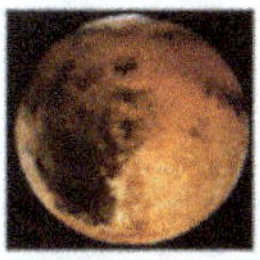	210°K (-63°C)	210°K (-63°C)	0°C
Earth 	278°K (15°C)	254.3°K (-18.7°C)	33.7°C
Venus 	737°K (464°C)	184.2°K (-88.8°C)	552.8°C

Fig 11. Radiation balance of the planets. For Mars and Venus, we use their measured albedos, and for the Earth the measured solar intensity. All are corrected for their orbital radii.

feedback processes produce nonlinear effects. There is radiation loss to space through the atmospheric window and the radiation from the various greenhouse gases originates from a fuzzy surface in the upper troposphere. But the Earth's energy balance is presently dominated by the combined effects of carbon dioxide and water vapor.

Earth is Unique

The Earth's atmospheric greenhouse effect is unique in our solar system. Mars has such a thin atmosphere (because its absence of a magnetic field probably resulted in its atmosphere being swept away by the solar wind) that its average surface temperature is determined precisely by the radiation balance calculation. It is cold! And Venus' temperature is dominated by a thick layer of that CO_2 greenhouse gas, (perhaps initially from all the H_2O that was evaporated from the surface; those molecules became the victims of photochemistry, with hot hydrogen escaping gravity). The discrepancy between the effective temperature for radiation balance and the measured surface temperature on Venus is over 500°C. Now that is some CO_2 greenhouse effect!! Can there be any doubt that all life on Earth, including humans and plants and animals, is favorably characterized by that special natural phenomenon of its very unique atmospheric greenhouse effect?

Global Warming

On Earth the initial source of atmospheric CO_2 was largely the oxidation (burning, for you non-chemists) of carbon-containing substances, with some additional contribution from volcanic activity and the slow decay of organic material from plants and animals. There have always been forest fires initiated by lightning, and CO_2 production increased after the discovery of fire by humans. These various atmospheric emissions were largely balanced by photosynthetic sequestration by plants and algae, and the conversion to calcium carbonate by ocean life and on land (limestone). Large amounts of CO_2 were also stored in the oceans in a steady state of solution capture and escape (just like in your sparkling wine). All of these processes were subject to temperature control from any outside influence on the radiation balance, this happened only occasionally in the distant past.

The increase in human population and the discovery and burning of fossil fuels have upset this equilibrium situation in the industrial age of the last century or so. We have greedily enjoyed this generation of cheap energy without regard for the Earth's environment and are rapidly increasing the atmospheric concentration of CO_2. The resultant enhancement of the atmospheric greenhouse effect has produced our measurable global warming. Actually the annual measurements of global surface temperatures frequently exhibit sometimes confusing weather-induced noise. We must examine the long-term trend increase of the global average of surface temperatures that is our evidence of climate change due to the greenhouse effect.

You protest, "So the greenhouse effect was in stable control of the climate in much of the past before the industrial age. What was the CO_2 concentration then?"

Hansen says it was about 350 parts per million, and there are some who think that should be our long-term goal. But the measurements show present concentrations above that and continuing to increase. In particular, there has been an increase of about 40% in the atmospheric concentration of CO_2 in recent years. This is attributed largely to the amount expected from the increased burning of fossil fuels. (The rate of CO_2 emission from volcanoes does not show the amounts and time variation to produce this result, and the photosynthetic and radioactive differences of the several carbon isotopes require the fossil fuel explanation.)

Atmospheric CO_2 concentrations have recently reached 400 parts per million, and the resultant slight extra CO_2 greenhouse trapping off-heat acts like a control knob on that thermostat to force an additional increase the evaporation of water vapor from the warming oceans. This initiates an increase in large positive feedback on the water vapor greenhouse gas concentration and we have increased trapping of the Earth radiation in the lower atmosphere. The average global surface temperature is increased slightly, but much of the increased energy now is also stored in the ocean or used to melt the ice of glaciers or that in polar regions.

Unfortunately these elevated greenhouse gas concentrations continue to trap the radiation above the –18°C energy balance altitude before a decreased amount escapes to space from the high altitude lower temperature and we no longer have energy balance. This energy imbalance has been carefully measured; the increase in greenhouse gas concentrations has disrupted the atmosphere's ability to balance the radiation budget. And we have global warming; the correlation of global surface temperatures with CO_2 concentrations has been documented.

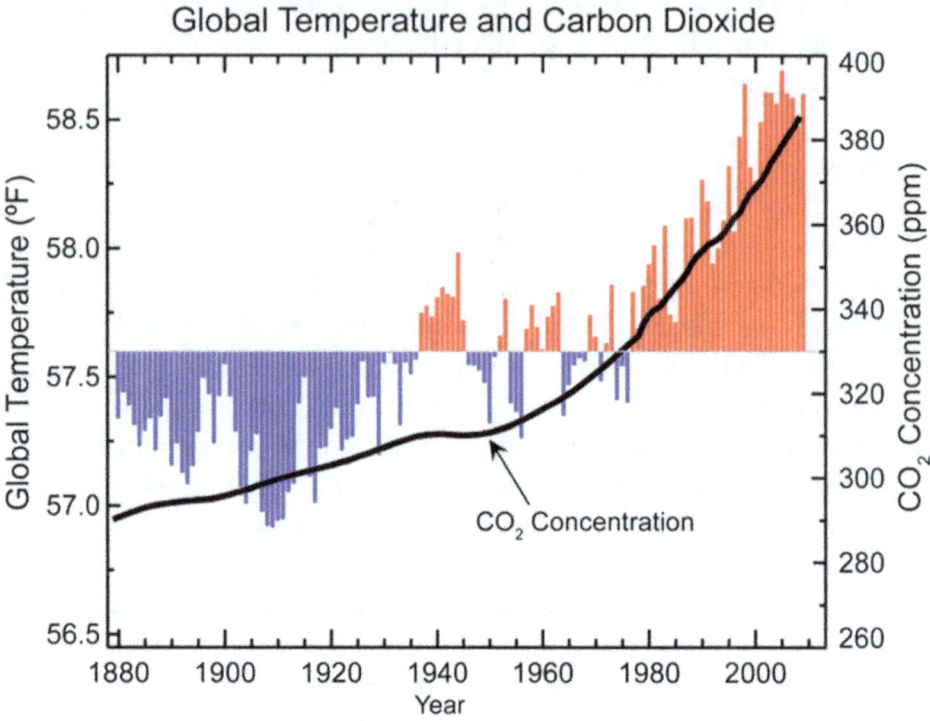

Global Temperature and Carbon Dioxide

Fig 12. Historical correlation of measurements of global surface temperature and atmospheric CO_2 concentrations.

In view of the failures of our society to move decisively in controlling the emission of CO_2, attention is being given to limiting the green-house effect of other greenhouse gases. Global warming potential is a term used to describe a molecule's importance in the atmospheric greenhouse effect. Methane concentrations are low but this gas is a much bigger molecule-to-molecule threat than CO_2. A decrease in the emissions of methane, which is comparatively quickly destroyed in the atmosphere by chemical reactions, could produce a prompt small warming reduction. Similarly, reductions in black carbon, tropospheric ozone, and nitrous oxide would be helpful; there are administration moves in these directions. But the present radiative forcing is the bottom line: CO_2 is the elephant in the room!

Atmospheric Lifetimes

You are probably aware that the Earth has experienced temperatures and climates quite different from the present. There were ice ages and there were dinosaurs at other times, for example. And some politicians will insist that is evidence that Nature is still in control. Ah yes, but we don't expect Nature to repeat certain of those behaviors, like tilting the Earth's axis of rotation, for a very, very long time. The situation has been relatively stable for much of recent human history, but with shorter cycles of atmospheric effects. You may have noticed that we have a cycle of water evaporation and precipitation; the period of the cycle, which we may call the atmospheric lifetime of water vapor, is about 10 days. Nature's water cycle guarantees that this by itself will not result in a runaway greenhouse effect. (Unless, of course, you are so stupid as to try to convert every molecule of fossil fuels into kilowatt-hours; then all bets are off!)

We also have carbon cycles, in which CO_2 from volcanoes and forest fires and decaying vegetation—or burning fossil fuels for energy—enters the atmosphere and is subsequently used in photosynthesis by plants or dissolved in the oceans. The atmospheric lifetime of CO_2 in these cycles varies greatly, from about 100 years for forest growth and decay to many centuries for other ocean and land processes. The CO_2 that we add to the atmosphere today will continue to trap heat in the atmosphere for a very long time. The 10-day water vapor cycle should follow right along with the CO_2 cycle because of that positive feedback. And of course you have noticed that there is another tricky little feedback: as the ocean is warmed, it can't hold quite so much dissolved CO_2, so that CO_2 escapes back into the atmosphere and gives an additional twitch to that control knob. Humanity's steady increase in the emission of the CO_2 greenhouse gas, aggravated by our ongoing destruction of the planet's forests, can be expected to disrupt the planet's stable radiation balance and produce a gradual change in the Earth's climate. And because of the very long environmental lifetime of CO_2, this climate change is irreversible during our lifetime.

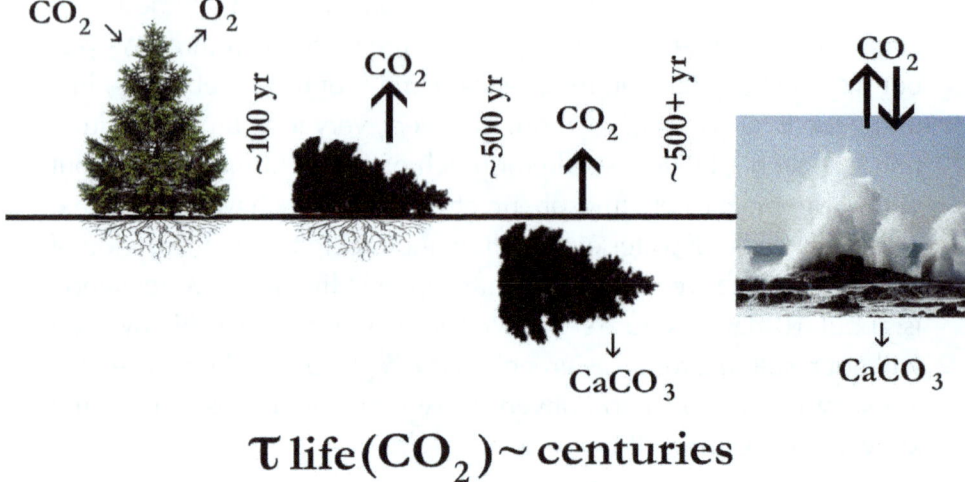

$$\tau \, \text{life}(CO_2) \sim \text{centuries}$$

Fig 13. Environmental lifetimes of CO_2 atmospheric cycles.

Methane and nitrous oxide are also increasing at rates comparable to carbon dioxide because of human activity, largely contributed by our agricultural practices. However, their lifetimes as greenhouse gases in the atmosphere are limited by chemical changes in the atmosphere. As we discussed earlier, methane is a more active absorber than carbon dioxide; fortunately its atmospheric concentration is much less and its atmospheric lifetime is only about 12 years. Each methane molecule contains 4 hydrogen atoms and one carbon atom. Its atmospheric chemical destruction yields carbon dioxide and ozone, and water of course, and these continue the greenhouse activity.

Methane is a principal constituent of natural gas and is particularly convenient for its use in electrical power generation. This is widely touted as the long-term solution for clean power generation since, when burned, it produces only half the carbon dioxide of coal. But remember that it unfortunately still produces half the amount of carbon dioxide as coal, and its extensive use in power generation continues to be a large source of atmospheric CO_2. It may continue to be valuable as a bridge source of power as wind and solar are developed and as a backup source for solar and wind variability, but it should not be relied upon as the ultimate solution for a stable climate. Unfortunately the procrastination in support of subsidies for wind and solar with the continued economic advantage of natural gas has led some industries to invest in long-term facilities in use of this fossil fuel. This constitutes a serious environmental mistake in the delay in control of climate change.

Nitrous oxide is produced largely in agriculture and has a lifetime of about 114 years. There is a practical limit to its reduction. While the present use of nitrogenous fertilizers in food production could be more efficient, the people of the world will always need food. Nitrous oxide is also the natural chemical source of the free radicals of nitrogen that control the amount of ultraviolet absorbing ozone—but that is a separate problem.

4

Earth's Response to Global Warming

A. The Physical Response

In Chapter III, we attempted to give a clear and complete description of the Earth's atmospheric greenhouse effect. The transport of energy by the continuous spectrum of heat radiation and the Earth's effective temperature for radiation balance are long established physics fundamentals. The large discrepancy of this radiation temperature with the measured surface temperature of the Earth clearly required an explanation for the heat trapping mechanism of the atmosphere. The theoretical details of this atmospheric greenhouse effect are thoroughly grounded in laboratory and atmospheric measurements. The comparisons with Mars and Venus demonstrate the control of Earth's surface temperature by the unique greenhouse effect of its atmosphere.

There is no question of the scientific understanding of this greenhouse effect control of Earth's surface temperature, critical for life as we know it. The new factor in the science is the increasing human population and our influence on the greenhouse gas parameters of the science. The measured increases in these atmospheric concentrations produce an addition in the heat trapping mechanism and the temperature increase of the Earth's surface, as required by science. Our global network of measurements, even with the uncertainties generated by weather variations and experimental methods, confirms the expected response. It is this measurement of global warming over an extended period of time that is the primary indicator of climate change. Earth's response to this warming demonstrates the additional climate effects driven by global warming.

I assume that you are aware that an increase of heat near 32°F (0°C) melts ice. You may not have been forced to learn that the phase change of ice to liquid water at that temperature requires a whopping 79 calories per gram, about the same energy it takes to warm liquid water three-fourths the distance from freezing to boiling. On Earth we always have a seasonal cycle of freezing and melting; now with global warming we should expect the balance to shift with decreased ice to increased liquid water. And with the albedo change from ice and snow to water and land, we expect another large feedback in global warming. This is particularly significant in the long Arctic summer, with the bright ocean ice changing to dark open water. The NASA and NOAA websites show that the Arctic Ocean ice has melted at an increased rate in the past 30 years, with the surface area observed by satellite in September 2012 decreasing from 7.5 to 4.7 million square km, a rate of 11.5 % decrease per decade, reaching its lowest area in 2012. As the Arctic surface feedback increases, we may expect increased loss of Greenland ice. In fact, satellite measurements of the ice mass on Greenland show a decrease of 100 billion tons per year; its ice loss doubled between 1996 and 2005. Antarctic ice is decreasing at 24 cubic miles per year since 2002. And we see the glaciers in our Glacier National Park and nearly all those in the rest of the world disappearing. The melting of Arctic Ocean ice, polar ice sheets, and the glaciers are clear indications of the increasing absorption of immense amounts of energy on the Earth's surface. And the phase change from ice to liquid water occurs without temperature change. So, the melting of polar ice sheets, Arctic Ocean ice, and the mountain glaciers are clear indications of the increasing absorption of immense amounts of energy on the Earth's surface. This absorption of heat in melting of the planet's ice acts as a sort of buffer on increasing temperature. So the observed temperature increase is that critical independent evidence of Earth's energy imbalance.

Fig 14. Greenland ice sheet meltwater.

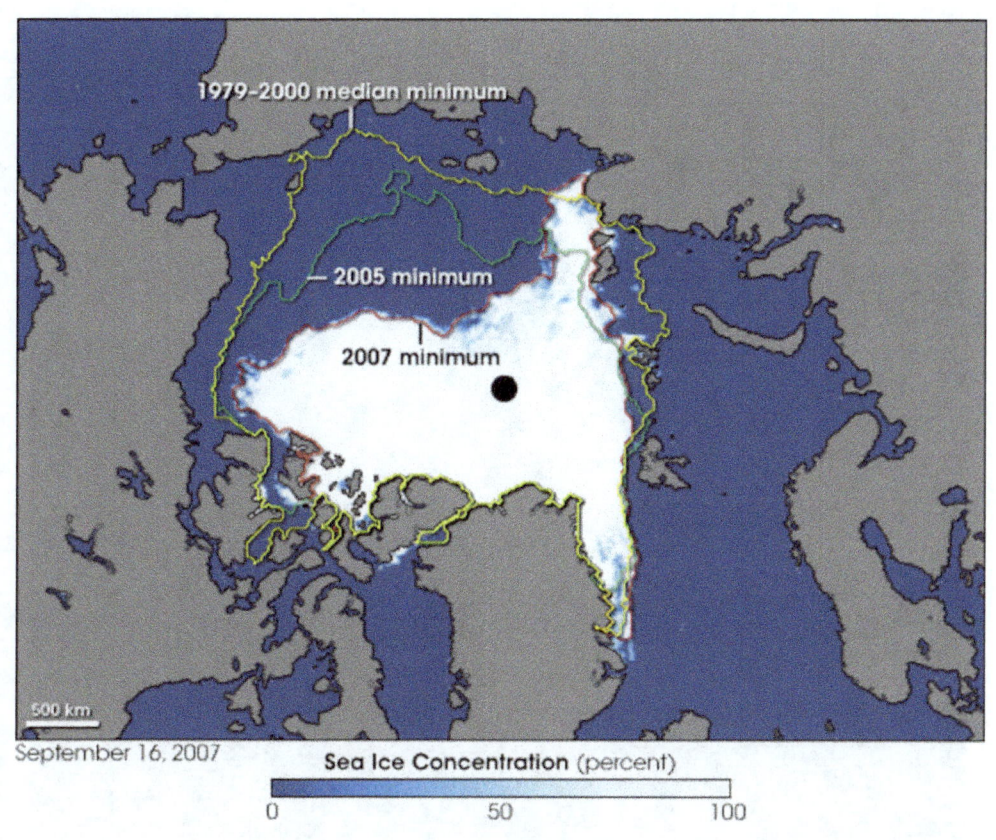

Fig 15. The arctic climate change is most spectacularly demonstrated in the summer melt of the arctic sea ice as shown. The minimum line of extent of the ice is shown in this graphic.

Fig. 16. Global Warming retreat of the Muir Glacier in southeast Alaska from 1941 to 2004.

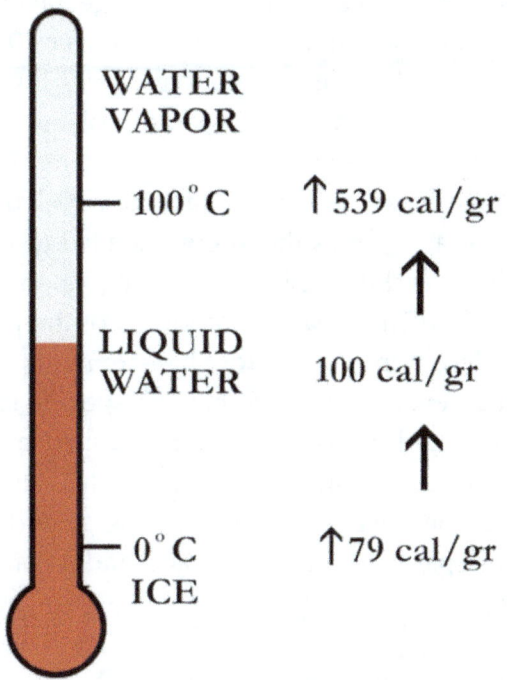

WATER
VAPOR

— 100°C   ↑ 539 cal/gr

↑

LIQUID
WATER   100 cal/gr

↑

— 0°C   ↑ 79 cal/gr
ICE

Fig 17. The large latent heat involved in the melting/freezing of ice that demonstrates our climate change, and the latent heat of evaporation/condensation of water that makes our weather so interesting.

Extreme Weather

There can be no doubt of the magnitude and direction of the above changes and the correlation with global warming caused by the increasing greenhouse effect. The IPCC and ongoing NASA and NOAA charts of average global temperature increase automatically suggest to the layman that the local response to global warming should be a gradual slight increase in daily temperatures. This can be managed with a slight adjustment of the thermostat. The result in mid-to-high latitudes is pleasant: less snow to shovel, more warm sunny days, and a longer and more productive growing season. But the IPCC report predicts increased frequency of extreme weather events.

So let us reexamine the weather science. A large fraction of the greenhouse energy is absorbed in the oceans that cover 70% of the Earth's surface. The increased radiation trapping warms the ocean and the evaporation gradually increases the large molecular water content and the water vapor energy in the atmosphere. The evaporation of water from the ocean surface requires a very large 539 calories per gram, more than 5 times the energy needed to heat water from freezing to boiling, and this energy is liberated along with the water during precipitation. The weather responses of the planet—extreme precipitation, wind, floods, and drought—are modifications of existing climate. These events do not come with specific labels of climate change, however, and are often mistakenly cited by contrarians as independent variations of a normal climate. It is the unpredictable sudden precipitation release of that extra energy and liquid water or cold snow that departs from that pleasant gradual progress to a new normal climate.

Citing these events as evidence of climate change is unconvincing to many; the planet's weather has always been subject to extremes. But it is inevitable that weather extremes will occur more often with the global warming increase in water evaporation from the Earth's oceans. Hurricanes, typhoons, and tropical storms; tornadoes, derechos, and haboobs; floods; drought; heat waves, stagnant air episodes, cold spells, and unseasonable warming and freezing are some of the events we can expect.

Still, the gradual temperature increase suggests that this erratic weather will be superimposed on a rather leisurely rate of climate change. Unfortunately a slow rate of increase will make these changes appear to new generations as their normal weather pattern; they will be unable to recognize the significant change from that experienced by earlier generations.

The particularly extreme weather events of tornadoes and hurricanes can lead to major local disruptions with considerable expense for repair, insurance outlay, and adaptation for future events. The increased latent energy of water vapor is a critical component of these events, but additional factors of atmospheric instability and circulation must be present. These storms can release great amounts of the global warming stored energy and precipitation whenever Nature arranges all the details. But heavy downpours and flooding must occur someplace sometime in any case. So why are these so variable and why do we have droughts and deserts?

May I remind you that the absorption of solar energy arriving perpendicular to the surface in the tropics is much greater than that spread over the same area in high latitudes. The Earth transports this excess energy to the polar regions by way of atmospheric and ocean circulations. Should we expect these to be changed by our increased atmospheric greenhouse effect? Let us look briefly at certain of our atmospheric and ocean circulations. A major part of the atmospheric transport is accomplished with the latent heat transport due to the evaporation and condensation of water vapor described earlier. The Hadley Cell, on the other hand, is the vertical convection and precipitation in the tropics followed by downward return of dry warm air at about 30 degrees latitude. Hadley Cell circulation has long been recognized as responsible to a great extent for certain of our low latitude deserts, the Sahara, Southwest USA, African Kalahari, and Australian Outback The increased energy of global warming is expected to increase the intensity and area of this circulation with expansion of the arid regions of the planet. And the El Niño and La Niña are notorious for affecting the world's precipitation as well as Pacific Ocean circulation. We must in the future look at a long-term average of these extreme weather events to document the validity of this aspect of climate change.

Ocean currents, notably the Atlantic Gulf Stream, transport much of the equator-to-polar energy. There is concern that the increased fresh meltwater and precipitation from climate change may eventually stall the thermohaline circulation of the warm surface water of the Gulf Stream with returning deep cold saline water from the Arctic, depriving European countries of their pleasant warmed climate. The periodicity and intensity of the Pacific circulation of the El Niño and La Niña may also be affected by global warming. And it is believed that currents of warm water under the seaward edge of Greenland and Antarctica glaciers have accelerated the motion of these glaciers into the oceans, with increasing sea level rise. These responses remain somewhat uncertain due to our early stages in the development of the scientific exploration of the vast oceans.

There are other climate components that leave no doubt of their connection to global warming. Let us look more closely at these effects even though they may not have occurred in your back yard. In contrast to the neglect of much of the public—I refer to your neighbors of different political persuasion across the street, of course—your government scientists have been busy collecting and attempting to communicate this climate data. These effects are accurately and elaborately documented on the Internet: the NASA website GLOBAL CLIMATE CHANGE-Vital Signs of the Planet and the NOAA website NOAA CLIMATE SERVICES. Perhaps you could take a break from social communications occasionally to check on the health of the planet. (I discovered these on the Internet, so you should have no problem finding them too. Or get your 5th grader to show you.)

Sea Level Rise

The temperature increase of the oceans on 70% of the Earth's surface is causing the familiar thermal expansion that contributes to sea level rise. This, along with the additional meltwater from glaciers and the polar ice caps, yields an average sea level rise of slightly more than 3 mm/year (a little more than an inch every decade). This may be increasing due to the more rapid polar glacier movements caused by the warmer ocean currents mentioned earlier. (The melting of the floating Arctic Ocean ice does not increase the sea level, however—check with Archimedes.)

There will be a significant increase in beach erosion with sea level rise, aggravated by possibly stronger and more frequent ocean storms.

We may unexpectedly lose our beach house and be forced to replace some of our coastal infrastructure: not a problem for vacationing coastal residents—we can adapt sometime in the next century. And some of our coastal cities, like Miami Beach at 4 ft mean sea level, may experience salt water street flooding at high tide. But there are also the folks living in Bangladesh who will have lost their livelihood, and the residents of the Maldives or some Pacific islands who must migrate to some significantly higher land. But hang in there—the extreme long-term effect of the melting ice is often quoted: melting all the Greenland ice would result in a 20 ft increase in sea level. Not to worry about this: in the interim other climate changes may have driven civilization to the brink.

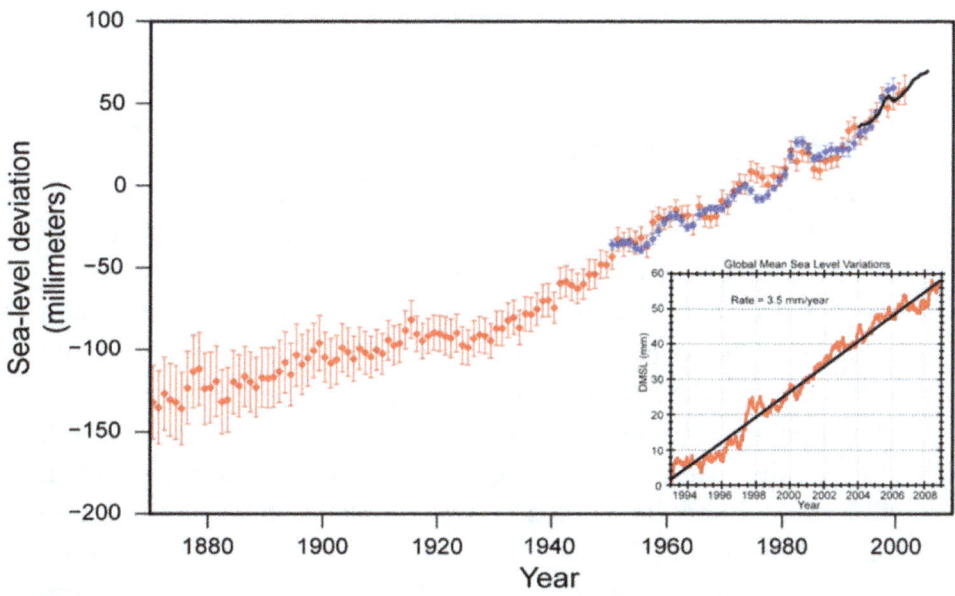

Fig 18. Measurements of sea level rise due to thermal expansion of ocean water and glacial fresh water runoff.

Many of the problems of climate change originate in the warmer oceans, and we cannot expect those warmer oceans to cool for centuries. This and the other natural environmental lifetimes are Nature's guarantee that the global warming that humans have initiated will not permit us to promptly return to yesterday's climate when we belatedly recognize and correct our violation of the planet's unique atmosphere. We did this by insisting on burning cheap fossil fuel and increasing atmospheric CO_2. What was the cost to the environment? Imagine spending the necessary funds to cool the oceans back to the previous norm. That would correct a very large part of the climate change difficulty. We owe a bunch! Payback is next to impossible. And Nature will increase the interest rate if we continue adding CO_2 to the atmosphere at the present rate. We have entered the Anthropocene. We should begin investing in clean energy to avoid increasing that debt.

B. The Biological Response

J. George is back. Professor Cook has demonstrated in the preceding discussions that the atmospheric greenhouse effect is responsible for the trapped heat at the Earth's surface that supports the plant and animal life of today. And we have seen that our increased human population is affecting the natural world around us. In particular, the human-caused increase in certain trace greenhouse gases like CO_2 is acting as a thermostat to enhance the water vapor greenhouse effect. It follows that we should experience additional global warming that will initiate other climate changes. We observe this warming trend in increasing average global temperatures and the accelerated melting of the planet's glaciers. Now Just George will carry on with the biological response of Earth to the changing climate and offer a few examples of what we are already experiencing and what we can expect in the near future.

In general we expect that most animal life, including humans, will have little difficulty in adapting to modest temperature increases. Many creatures will have the ability to migrate the necessary distances to function normally. Humans have developed technologies of heating and air conditioning that have expanded the temperature range of comfortable activity, although man cannot quickly adapt to the trials of working outside in much higher temperatures. While a modest temperature increase may be tolerated by humans and possibly be viewed as desirable, the responses of some of the more sensitive creatures and plants should be examined. For example, the Rosy Finches of the Rockies nest above tree line near the Continental Divide and retreat to lower elevations each winter. In spring, large flocks gather at these lower elevations and gorge themselves in preparation to migrate. But in recent years these gatherings have not happened. Observers have seen no Rosy Finches. The change is attributed to the fact that at nesting time they search for food on open snowfields, which have disappeared due to climate warming. The birds have been observed in smaller numbers some distance to the north. Another example: pikas are severely threatened. These cute little egg-shaped round-eared relatives of the rabbit live in burrows in high altitude talus fields convenient to alpine meadows. Because they do not hibernate, pikas must cut and store grasses to provide winter fodder. As the Earth's climate changes, high altitudes are becoming warmer in summer and winter snowpack is decreasing. Pikas are very sensitive to heat (as little as six hours at 77°F or above is lethal) and control their body temperature by retreating to cool burrows under rocks. The rise in ambient temperature limits the time they can spend in the open collecting food. Availability of their favorite grasses is changing as the climate warms. And because of the decrease in winter snowpack, burrows are not as well insulated and pikas must eat more to maintain body heat. All of these factors may contribute to the decline of the species. Does that matter to you? Maybe not, unless you are another pika, or a mid-size mammal or raptor who depends on pikas for food. But any loss of biodiversity leaves humans poorer for the loss. Monitoring the health of the pika population and their numbers will be an indicator of the severity of climate change our planet is experiencing and an early warning to the human population.

The temperature response of Earth's plant life is much different with a more restricted range characterized by each fixed location. And of course, all life on Earth has evolved to adapt to the daily and seasonal ranges of temperature. Much of the life on Earth has a very narrow range of success. When we depart from that, even humans will suffer the ability to adapt. The support system for animal life is greatly dependent on the well-being of plant life, and life support for humans depends on both plants and animals. All life depends on the ready availability of water. Temperature change has a direct effect on the evaporation and precipitation of the water cycle. So we see the complex manner in which temperature change controls all aspects of the climate, and ultimately human life.

We should use caution in thinking we can evolve quickly enough to adapt to the changes of the future: recent studies with bacteria have shown that it takes some 2,000 generations to develop new complex traits.

Significant losses in biodiversity have already been observed in some ecologically rich sites worldwide. Fragile ecosystems and fragile species are in extreme danger of extinction. Botanical organisms have definite geographical limitations, based on temperature, soil salinity, light, water, ambient weather, and latitude: bananas don't grow in Iowa and wheat doesn't grow in Costa Rica. The normal range of plants we grow for food, clothing, shelter, and fuel will move toward the poles as the climate changes. For example, the deep black soils of the American Midwest have been responsible for producing immense crops of wheat and corn. So with global warming, can farmers switch to growing cotton? And if food crops don't produce as well in the thin soils of northern Minnesota and Wisconsin perhaps some plant breeder can improve their DNA. As climate change progresses, crops in Kansas and Nebraska will produce diminished yields because of higher heat and drought conditions. Wheat will grow farther north, although it is unlikely it will ever equal current levels of production, and it will lead to destruction of northern forests. Our nomadic ancestors could respond to deteriorating conditions by moving on; we must consider the additional costs relating to the availability of markets, locations of processing facilities and the transportation infrastructure.

Protein produced by animals, especially cattle, is expensive, both in cost and carbon footprint. As more of the Earth's forests are sacrificed for use as rangeland, this will become more critical. Death of coral reefs will severely impact fisheries.

Bark beetles, twig borers, white pine blister rust, and many other insects and diseases are increasing in forests worldwide. Trees stressed by drought are more susceptible to invasion by pathogens and dead trees increase forest fire danger. Diseased trees produce significant amounts of methane in their dead inner cores, according to a recent Yale study, multiplying the fire risk. The spring pollen season in the US is occurring earlier and lasting longer due to climate change. Ragweed, an aggressively invasive plant that produces increased amounts of highly allergenic pollen in response to increased temperatures and carbon dioxide, is spreading. Recent dry seasons in almost all citrus growing regions of the world have led to increased numbers of Asian citrus psyllids, a tiny insect that transmits a bacterium responsible for huanglongbing (sometimes called citrus greening), reducing edible production and resulting in the forced removal of millions of citrus trees. Rising temperatures and recent drier-than-normal wet seasons in Central America have exacerbated the spread of coffee rust, resulting in up to 70% crop losses for marginal small growers who can ill afford a failed season.

Changes in temperature and precipitation, as well as droughts and floods, affect agricultural yields and production. Although climate deniers suggest that increased atmospheric CO_2 and warmer temperatures will increase plant growth and food for a hungry world, the situation is not quite that simple. Increased temperatures affect photosynthesis by denaturing the enzymes necessary for production of sugars. Plant respiration rates accelerate with temperature rise, sometimes resulting in more energy used than produced. Pollen is adversely affected by higher temperatures; corn pollen dies at about 38°C. In some regions of the world, changes in temperature and precipitation may compromise food security and threaten human health through malnutrition, the spread of infectious diseases, and food poisoning. In other parts of the world, important species may become endangered or extinct.

Drought affects the availability of clean water, concentrates contaminants negatively affecting surface waters, and reduces agricultural productivity, resulting in increased food prices and food shortages. Access to clean drinking water is already a huge worldwide problem. It is estimated that by 2050 between 350 million and 600 million people in Africa and up to a billion in Asia will experience increased water stress due to climate change. Fifty years ago the common perception was that water was an infinite resource. There were fewer people on Earth and they were not as wealthy as they are today, they consumed fewer calories and ate less meat, so only a third of the volume of water we need today was needed to produce their food. Now, with more than seven billion people on the planet, the consumption of water-thirsty meat and vegetables is rising and there is increasing competition for water from industry, urbanization and production of biofuel crops. The total amount of available freshwater is decreasing because of receding glaciers, reduced stream and river flow, and shrinking lakes. Many aquifers have been exhausted and are not recharging quickly. Although the total fresh water supply is not used up, much has become polluted and contaminated, rendering it unsuitable or otherwise unavailable for personal use, industry and agriculture. To avoid a global water crisis, farmers will have to strive to increase productivity to meet growing demands for food, while industry, cities, and individuals must find ways to use water more efficiently.

(Actor Alan Alda has warned us of excessive use of scientific jargon in his recent attempts to promote better science education. But most of the following microscopic bugs do not have nice common names like robin or bluebird or dandelion. I am professionally familiar with these guys, so please try to give token attention to this next paragraph and move on to my added indications of how nasty they are.)

Climate change will cause higher air temperatures, increasing cases of food poisoning caused by *Campylobacter, Salmonella, Shigella,* and *Escherichia coli*, especially in parts of the world with inadequate refrigeration. Flooding from severe storms may cause sewage treatment plant overflows, contaminating crops with those same pathogens as well as increasing the occurrence of *Listeria* contamination. Boulder authorities warned residents returning to their homes after the 2013 flood not to eat anything salvaged from their gardens: floodwaters had overrun water treatment plants, releasing *E. coli* into waters that were also contaminated with industrial and agricultural chemicals and petroleum products from damaged and overturned storage tanks. In some areas of the third world, primitive sanitation facilities can be overcome by heavy precipitation during the rainy season, sometimes flooding wells and contaminating water supplies. Heavy rainfall can prompt an increase and wider dispersion of *Cryptosporidium* and *Giardia*. Reservoirs and lakes in California were reported to host an increase in leech populations in summer 2012. An amoeba (*Naegleria fowleri*) that causes a type of encephalitis appears to be migrating northward due to increased temperatures. The soil fungus *Coccidiodes immitis,* common in arid regions of North and Central America, produces spores when the rainy season begins. The spores become airborne when soil is disturbed and cause a respiratory illness called Valley Fever when inhaled. Recent outbreaks among workers at desert photovoltaic facilities in California and Arizona have been noted. In sub-Saharan and West Africa, the spread of meningococcal (epidemic) meningitis is linked to drought. Cases of shistosomiasis have increased. Blue-green algae blooms in lakes are unsightly, smelly, and can cause illness in humans and pets, as well as killing fish and other aquatic organisms by depleting oxygen supplies. Toxins produced by an algae bloom in Lake Erie in summer 2014 forced officials to shut down Toledo, Ohio's, water system. Agricultural runoff is largely to blame for these algal events. Ocean algal blooms produce potent neurotoxins that are often taken up and bioaccumulated in filter-feeding mollusks, including oysters, clams, and mussels, as well as by certain marine and freshwater fish. Health effects for humans ingesting affected seafood include amnesia, diarrhea, numbness, liver damage, skin and eye irritation, respiratory paralysis, and other symptoms that may be severe, chronic, and may even lead to death. It has recently been reported that even a single low-level exposure to algal toxins can result in physiological changes indicative of neurodegeneration. Hundreds of Florida manatees have died recently due to toxins produced by algal blooms. And it was recently reported that giant viruses, almost as large as bacteria, have revived after being frozen in permafrost for 30,000 years. Fortunately those observed (so far) have only infected amoebae.

Mosquitoes are expanding their geographical habitats. *Aedes aegypti* has returned to the San Francisco Bay Area after a forty year absence. Species traditionally limited to the tropics are now in the Florida Keys and continuing their northward exodus, bringing with them malaria, dengue fever, chikungunya, and other diseases previously confined to the tropics. A surprisingly widespread outbreak of dengue fever affected Tokyo, Japan, during summer 2014. Chikungunya, a painful but rarely fatal viral disease with no preventative measures or treatment, was first reported in Haiti in Spring 2014. Within a week there were 1500 cases. The disease has since spread to Central America (where over a million cases were reported in the first month) and throughout the South Pacific. Outbreaks of West Nile virus, also mosquito-borne, were reported in all 48 lower states and were particularly severe in Texas in summer 2012. Ticks carrying Lyme disease are expanding their territory northward. Cases of hantavirus, transmitted by deer mice, and bubonic plague, transmitted by fleas on infected rodents, have increased in the southwestern US.

Some problems will simply be new for new geographical areas; plant breeders have already solved some of the problems climate change will pose. Researchers in Australia have developed a wheat variety more tolerant of high temperatures. Scientists in Florida have developed tomato varieties tolerant of heat and humidity. British researchers have had some success with rice varieties tolerant of higher salinity, developed after the 2011 Japanese tsunami sent sea water over thousands of acres of farmland. Israeli scientists have bred barley varieties tolerant of salt water, grown also in California's Sacramento delta where increased draw-off of water from the Sacramento-American-Feather

River systems has led to further salt water encroachment from San Francisco Bay, rendering formerly fertile farmland unsuitable for row crops. In some cases, increased labor can ameliorate problems: seedlings of crops tolerant of alkaline conditions can encounter an impenetrable crust on the soil surface which must be mechanically broken up or softened with (often unavailable) water; if a scarcity of water prevents flooding of rice paddies (a practice utilized mainly for weed control), some other means of combating weeds must be found. At the same time, a common herbicide utilized extensively for weed control appears to be contributing to the demise of Monarch butterflies by eliminating milkweed, necessary in the development cycle of Monarchs. These are difficulties faced in developed countries. Subsistence farming in third world countries will suffer even more. Families faced with starvation will be forced to migrate or die.

Not only will disease organisms and insects thrive, there will also be related, if predictable, difficulties. For example, stored grain and nut crops are particularly susceptible to infection by *Aspergillus flavus,* a common soil fungus, if they have been stressed by drought. Aflatoxins produced by the fungus can cause stunted growth and delayed development in children and cirrhosis and liver cancer in humans of all ages. In summer 2012, Midwest farmers harvested their drought-stressed unmarketable corn, chopping it for silage so it would not be a total loss. When fed to cows, aflatoxins in stored silage can be passed on in milk, prompting Iowa health officials to institute mandatory screening of milk samples.

Although the European Environment Agency has stated that biofuels do not address global warming concerns, increases in oil prices and a perceived need for energy security has prompted an increase in bio-fuel production in the last ten years. In the US, corn is the usual raw material used in ethanol production. However, the cost of the total energy consumed by farm equipment in cultivation and planting, the fertilizers, pesticides, herbicides, and fungicides made from petroleum, capital cost and maintenance of irrigation systems, harvesting, transportation to processing plants, fermentation, distillation, and drying, transportation to fuel terminals and retail pumps and the lower energy content of ethanol fuels combine to deliver little financial or energy savings for the consumer. In addition, the diversion of food crops for energy production has the potential to create additional food insecurity in marginal communities. (The increased price of corn in northern Mexico recently created a tortilla shortage in the area.) Solid biofuels (wood, sawdust, domestic refuse, agricultural waste, dried manure) are often a waste product of other processes, which eliminates the competition between food and fuel; however, combustion of solid biofuels emits high amounts of particulates and hydrocarbons.

Climate change will lead to more frequent, more severe, and longer heat waves in summer, as well as more episodes of stagnant air. Record drought afflicted Australia heat in summers 2013 and 2014, as well as record heat which caused the deaths of hundreds of thousands of bats and flying foxes resulting in public health concerns. Heat waves increase the demand for electricity to run air conditioning, which in turn increases air pollution and greenhouse gas emissions from power plants. Heat increases ground-level ozone, formed when nitrogen oxide, carbon monoxide and volatile organic compounds (products of vehicle exhaust, chemical solvents and industrial emissions) react in sunlight. Increased concentrations of ozone can irritate the respiratory system, reduce lung function, aggravate asthma, and increase susceptibility to respiratory infections. This situation is especially harmful to young children, older adults, outdoor workers, and those with asthma and other chronic lung diseases. This tropospheric ozone adds insult to injury in behaving like a greenhouse gas when it absorbs Earth's infrared and shares the energy with the lower atmosphere. Airborne particulate matter, originating from power plants, smelters, wood combustion, and forest fires, can also exacerbate respiratory health issues.

Warming of the oceans, affecting weather patterns and global ocean circulation, and acidification of the oceans will alter marine populations resulting in disruption of fisheries and not only economic fallout but also food insecurity. Warming oceans have been implicated in the decreased amount of marine fog experienced on the Northern California coast, impacting *Sequoia sempervirens*, the famous coast redwood, which is dependent on coastal fog as a water source. Winter tule fog, a radiation fog directly dependent on rainfall, in California's Central Valley has decreased 46% over the last 32 years, reducing chill periods necessary for some crops. On land, species are straying from their native habitats at an unprecedented rate: 11 miles toward the poles per decade. Areas where temperature is increasing the most show the most movement by native organisms. Migration of disease vectors (organisms that do not cause disease but transmit infection by carrying pathogens from one host to another) and animal hosts that carry certain diseases will accelerate.

Virtually the whole world has seen an increase in wildfires in the last few years. Major wildfires in Russia and Europe have cost lives and resources; events in Argentina and Australia wreaked similar damage. In the western US, fires in Colorado, Washington, Oregon, California, Arizona, Idaho, Utah, Texas, Oklahoma, and New Mexico have taken lives and property, destroyed biomass, cost staggering amounts to fight, produced dangerous levels of unhealthy smoke, and contributed to increased levels of pollution both locally and globally. Drought in California has extended the wildfire season into January 2014, a time of year that should see adequate rainfall to prevent wildfires. The firefighting and property damage does not include the increase in atmospheric CO_2 and the decrease in forest photosynthesis sequestration of that greenhouse gas. These forest wildfires are illustrative of the increasing costs of adaptation to climate change.

One and a half billion people currently live in coastal areas that are vulnerable to sea level rise and the hazards of inundation, storm surge, coastal erosion, and the encroachment of salt water on fresh water supplies. Beach erosion and coral bleaching are affecting fisheries as well as the value of tourism destinations. Elevated soil salinity lowers agricultural productivity. Bangladesh, India, and many small island states such as the Maldives and Tuvalu may have to relocate large populations over the next 50 years as sea levels rise, as many as 80 million people from Bangladesh alone. Fifteen of the world's twenty megacities (cities with populations over 10 million) are in danger from sea level rise. The sheer numbers of evacuees will challenge immigration laws worldwide.

Melting glaciers result in loss of a water source utilized for personal and agricultural use and power generation, as well as risks of flooding and rock avalanches from destabilized slopes. Black carbon (from fossil fuel transportation, forest fires, agricultural burning, and cooking fires) and dust deposits on ice fields accelerate the melting rates. Native fishing villages in Alaska are literally falling into the sea because of the instability of thawing permafrost and its susceptibility to erosion. Archeological artifacts revealed by thawing permafrost have been destroyed by storm periods lasting weeks longer than "normal" because of lost ice cover. Thawing permafrost is also a source of methane, along with cattle, rice fields, and any uncombusted emissions from natural gas wells.

Extreme weather events, sea-level rise, destruction of local economies, resource scarcity, and associated conflict due to climate change are predicted to displace up to 200 million people worldwide by 2050. Experts in the fields of social and behavioral health anticipate an increasing incidence of stress-related maladies due to displacement and involuntary migration, while nostalgia for "the good old days" may lead to depression.

Now I will share computer access with Professor Cook so we can describe how we Earthlings have been responding to climate change and what we should be doing.

C. The Human Response
Global warming is the straightforward natural response to the

enhanced greenhouse effect resulting from our increased emission of greenhouse gases. (The adjective "enhanced" is unfortunate because it implies that it's a good and desirable thing. There can be too much of a good thing!) We have described other climate changes taking place. Some of our neighbors are beginning to notice; these things just happen, they always have, always will. These mild winters are welcome; summer heat can be managed with a little bump of the thermostat on the AC. We will adapt and technology will find solutions to whatever happens. And if it gets really bad the government will step in and fix it for us. But don't you wonder if this plan will work for more than a few years of continually increasing greenhouse gas emissions and the inescapable global warming?

Declining sea ice coverage will improve access to arctic waters. In the case the Arctic Ocean, increased access means that issues of sovereignty (priority in control over an area), security (responsibility for policing the passageways), environmental protection (control of ship-based air and water pollution, noise, or ship strikes of whales), and safety (responsibility for rescue and response) will become international political issues. As the Arctic ice melts, the world is becoming more aware of the resources below. According to the U.S. Geological Survey, 30 percent of the world's undiscovered natural gas and 13 percent of its undiscovered oil are under this region. As a result, military action in the Arctic is heating up, with the United States, Russia, Denmark, Finland, Norway, Iceland, Sweden and Canada holding talks about regional security and border issues. Several nations, including the US, are training troops in the far north, preparing for increased border patrol and disaster response efforts in a busier Arctic.

Of course there may be good results from disruptive climate change. Extreme weather events can be isolated, some bad, some good. Adverse conditions might lead us to discover new sources of food, and we might learn to be accepting of the desolate results of drought, flood, or windstorms. The best result of all would be a cooperative response by all societies in recognizing climate change as a global adversary.

But events suggest that this will be extremely unlikely. In a report solicited by the White House in the early 1990's, the Pentagon recognized climate change as the number one danger facing the American people. Pentagon experts cited changes in global temperatures as the potential cause of civil and political unrest, mass migration brought about by environmental degradation, the weakening of national governments, pandemics, and immigration stress. Even in the U.S., while many of our citizens work to generate clean energy, it appears that a large section of the population seems hell-bent on destroying civilization by burning coal, oil, and gas.

And the argument continues to be made by certain voices in our country's establishment that it would be futile to make economic sacrifices in our corrections of greenhouse gas emissions while the exploding populations of China and India continue to increase the global atmospheric pollution. I wonder; have we abandoned our role in global leadership?

It has been said that we have a choice. We can work to Manage the Unavoidable. That is, we can adapt to the disruptive climate change that occurs after we have continued business as usual with uncontrolled greenhouse gas emissions. Or we can work to Avoid the Unmanageable. That is, we can mitigate the future climate disasters by correcting our mistakes in energy production and transportation caused by reliance on use of fossil fuels, by developing the solar, wind, or nuclear energy sources of zero carbon emission. This won't return us to our normal climate but will make the results less disruptive.

But let us be clear that the additional slight mitigation steps that we might take in advance of climate disaster will not avoid the extreme adverse effects of business as usual. For example, we might practice some energy conservation in building construction and auto mileage improvements that would delay the inevitable. Or we could strengthen our protective beach sand dunes to postpone flooding of our parking garage. Or we might restore our Florida water management to flood the Everglades with fresh water that might slow the future en-croachment of saltwater. Or we might construct a defensible space about our mountain home that could save it from all but the most monstrous future wildfires. But it will require the major mitigation process of a prompt transition to low carbon energy production to prevent unmanageable climate change.

Adaptation

If we continue on as we are, pouring vast amounts of CO_2 into the atmosphere and forcing Earth to change, our descendents will have to adapt by changing food production, transport, and energy genera-tion. Hopefully they will have the intelligence to modify their human constructs of governments and economics to live in harmony with Nature. Adapting to the rising seas and extreme weather will be a long and increasingly costly business. And Nature is not going to re-spond promptly when your Congressman finally gets the message to get the greenhouse gas emissions under control. While you sit back and twiddle your thumbs as big business increases the greenhouse gas pollution, Nature will gradually depart from the stable atmo-spheric situation of the past to produce a climate change that will be extremely slow to return to normal. If we finally get around to correct-ing our increased emission of CO_2, Nature will not quickly return us to the normal climate of yesteryear. And it's going to cost you!!

Already a few of our fixed income folks are having problems coping with triple digit summer temperatures and rising food costs. And a few of our budget managers are beginning to worry about the in-creasing costs of fighting wildfires and repairing tornado damage. An indirect universal consequence will be increasing insurance rates for all of us. The financial costs of adaptation are increasing!

Oh, but perhaps these are temporary climate changes from some hitherto unforeseen cause. But we all buy fire insurance to cope with unforeseen calamities (except for those folks living in the forested foot-hills of Colorado who can easily foresee potential disasters). In fact, our rules of economics force banks to require insurance to qualify for a mortgage. A similar risk assessment that recognizes the adaptation costs of fighting western wildfires, repairing Colorado flood damage, and the costs of other extreme weather events such as Superstorm Sandy or the Philippines super typhoon Haiyan should override any uncertainties about the human cause. The climate folks have adopted this idea and call it the precautionary principle. The sensible procedure is to invest in methods to mitigate the effects of global warming, even if there is the slightest doubt of the nasty consequences. A single homeowner may rationally argue that his solitary efforts to diminish his carbon footprint with solar panels or improved insulation are of little consequence (even though that investment may reap a better return than leaving the cash in a bank savings account). Hopefully there will be an increase in the few corporate investors who see the moral need and long term advantage in pursuing this course.

But in a few quarters there is an increasing awareness of undesirable consequences of climate change. The Gold Coast communities of South Florida are gearing up to adapt to rising sea levels. Beach erosion will increase with a few more millimeters of sea level and the likely aggravation of stronger and more frequent Atlantic storms. So we must expect to continue to protect the condo investments and tourist accommodations with local taxes, and lobby for continued Federal support for beach replenishment with diminishing off shore sand. And saltwater intrusion of fresh water aquifers has always been an imminent threat. The inland fresh water impoundments were installed in response to the ill-advised attempts to drain the Everglades. Must we now raise the levees? So Floridians are developing a master plan to deal with salt-water intrusion. And they are also searching frantically for funds to restore the Everglades from past mistakes. But there will be salt-water inundation of Everglades National Park well before it gets unpolluted fresh water from this admirable plan.

But look, stupid: the sea levels are rising because of that temperature increase, and that's because you continue to dump CO_2 into the atmosphere from all those Florida coal fired power plants (77 million tons of CO_2 in 2000). And we happen to personally know of a solar power engineer in Boca Raton, Florida, who could help you postpone those problems for a while. Of course the TV executives have assumed no responsibility for communicating news of climate change; they have to maximize profits from the fossil fuel commercials. It would be good to educate those retirees who vote, but how do you get them to shut off that TV garbage? The climate may change only slightly during their remaining years; there is a more urgent precautionary principle that requires concentration on church attendance and the collection plate rather than the greenhouse effect and solar panels. Perhaps they have also abdicated any responsibility for the welfare of their children and grandchildren? Florida is going down the tube—or the sea level equivalent!

Hurricane Katrina in New Orleans was a wakeup call for attention to the dangers of extreme weather. There has been speculation that the landfall of a category 5 hurricane on New York City might spur early attention to the need to curtail the increased emission of human production of greenhouse gases. Hurricane Sandy, the very destructive hybrid storm of October 2012, was a prelude to this stimulus, and winter storm Nemo added evidence by demonstrating the effects of added water vapor from warmer oceans. Will we continue to ignore this danger of human-caused climate change?

A History Lesson for Today

There is a history lesson pertinent to today's problems of global warming and climate change. Only the over-30 generation will have personal recall of the environmental danger of ultraviolet exposure due to stratospheric ozone destruction, but all of human civilization has benefited from the prompt and successful international resolution of this problem. Twenty-five years ago we discovered the destruction of UV-absorbing stratospheric ozone by man-made chemicals, which threatened to expose every person on Earth to increased solar ultraviolet and risk of skin cancer. Scientists had discovered that minute amounts of certain nitrogen compounds were important in the natural balance of our protective ozone layer. When the proposal for subsidizing the aircraft industry in the development of supersonic transport was raised, scientists sounded the alarm about the resultant stratospheric pollution increase and a dangerous mistake was avoided. With the scientific discovery of a similar danger from chlorine compounds, the chemical industry protested the loss of jobs and profits. The danger seemed uncertain and proposed adaptations were dark glasses, protective clothing and sunscreen for humans. Plants and animals would remain vulnerable. Environmentally concerned citizens refused to purchase spray cans of shaving cream and cosmetics powered with CFCs. The reality and full danger of this imminent disaster became apparent when scientists observed complete destruction of the ozone layer over springtime Antarctica, the spectacular ozone holes that still reappear. Prompt action by the international community led to the Montreal Protocol restrictions on production and use of the damaging CFC compounds. The atmospheric lifetime of these molecules is very long, however, and the chemical reactions will continue to produce the ozone holes for another 50 years or so.

The activities of our increased human population have now generated a problem of global warming and climate change. The problem and its solution are more complex and difficult than those faced with the ozone hole, and the dangers are more general. Human health, shelter, and food are jeopardized; adaptations will be difficult and costly. Our life support system of plants and other animals may ultimately change or cease to exist. The solutions are again in conflict with our political system and economic ambitions.

The fundamental science of the atmospheric greenhouse effect is without question. However, a recent poll indicates a confused public: 34% mistakenly believe climate change is due to the ozone holes. In fact, ozone and CFC are greenhouse gases similar in action to carbon dioxide, methane, and nitrous oxide, but only in the lower atmosphere where the absorbed infrared energy from the Earth can be reradiated to the ground to form our warm radiation blanket.

One might hope that the international response to climate change would follow the example of the successful resolution of the ozone problem. Actually the economic response is similar. The industries at fault work to protect their profits with misinformation and warnings of damage to business as usual. Unfortunately the danger is not so uniformly apparent to all societies. We have to wonder what the continuing dominance of politics and economy with business as usual will do to our future climate. Mother Nature will not turn off the atmospheric greenhouse effect; she will increase the global warming in response to every molecule of CO_2 that we add to the atmosphere from the frantic burning of fossil fuels. The planet will follow the warming with increased extreme weather and sea level damage with burgeoning adaptation costs. On the other hand, with enlightened governance our country could be our planet's leader in solving this problem. But if there is no political change, the economy will suffer with the burden of increasing adaptation costs. If civilization is lucky, the industrial activity might eventually settle in to the activities of the Middle Ages with the corresponding reduced CO_2 emissions. And a few centuries later Nature will sequester the extra CO_2 and we can start over, perhaps a little wiser.

Mitigation

The most immediately effective and least painful procedure is to conserve energy and produce food more efficiently. Our commercial, government, and residential buildings lose heat energy at an exorbitant rate. This could be corrected with improved insulation in existing and newly constructed buildings. Perhaps you should think of this as your personal recognition of the precautionary principle. Besides, in one of the very few TV programs on climate change, we learned that replacing all the Empire State building windows was repaid in just four years of reduced energy costs. It is of course true that individuals who conserve energy with improved insulation or who invest in geothermal heat and cooling in new construction will have little influence on emissions; those corporations who have a monopoly on transport and electricity continue to produce cheap energy for the rest of society because they do not pay for the increased CO_2 greenhouse effect with its global warming and climate change. And the rest of society pays for the increasing costs of adaptation to the climate change.

Our ancestors generated heat for food preparation and comfort with use of biofuels—burning wood—a net zero carbon emission procedure. The CO_2 produced was balanced by the photosynthetic trapping of CO_2 in growth of new forests. The fireside source of heat continues to be necessary for populations in developing countries and popular for residents near sources of firewood from forest management activities. We have also expanded this net zero carbon procedure for farm crops in production of ethanol for transport fuel. But this has initiated a serious competition for food production, particularly serious for the world's poorer societies. Clearly this is a limited solution for reducing CO_2 emissions.

The ultimate solution requires a major reduction in CO_2 emissions in production of dwelling heat, fuel for transport, and generation of electric power. There is an immense alternative energy source at hand in direct solar radiation and indirect wind power. Implementing technological advances in harnessing a very small fraction of this power would halt further increases in global warming. This source is frequently criticized as being intermittent and undependable. In reply, very large wind farms in open areas of persistent wind or solar installations in sunny deserts are being constructed, but many require additional transmission lines to deliver the power to the cities. A possible solution of residential or small business solar installations extending uniformly over a large area could alleviate the intermittency difficulty and utilize the existing power grid for transmission. However, individual owners must be convinced of the global warming danger and forego an existing convenient cheap power supply. They are unlikely to uniformly invest in carbon-free energy without government incentives or subsidies. Only with an informed and dedicated electorate is this likely to change.

While we are busy making excuses for our apathy and sluggishness in reducing emissions, Germany, a country about 1/27th the size of the US that gets about as much sunshine as Alaska, is third in the world in the production of alternative energy generated electricity, currently producing five times as much as is produced in the US. Their *Energiewende*, the German plan to end the use of fossil fuels and nuclear power, has the support of all political parties and is on track to achieve the goal of using only clean renewable solar, wind, and biomass generation of power by 2050. Instead of looking at "problems" that prevent implementation, they look at "tasks" to be analyzed and resolved. What disagreements there may be are not over whether anthropogenic climate change is taking place, but rather how to speed the transition. Approximately 65% of the new carbon-free infrastructure is owned and controlled by individuals and communities.

Geothermal heat has always been a practical source of power in certain very active regions. Now it has being rediscovered as an efficient source of home heat with relief of much of the carbon producing electric power. And some progress is occurring in personal transport with the development of electric cars. However, decreased greenhouse gas emission will only occur with non-fossil fuel generated electric supply. We could also expand our energy production from nuclear fission; it is a zero CO_2 production process. There are certain inconveniences and dangers in its use, but these are clearly solvable with proper technology. The "not in my backyard syndrome" has delayed its implementation and this energy source must be supplemented by the more available methods described above.

Or we can continue to dream of the unlimited energy from nuclear fusion—just like on the sun but controlled in our laboratories. We hoped for this to happen 50 years ago when this physics professor helped measure the high temperatures and stability of the hot plasma. Now we have costly local and international research facilities with endless optimism to control those instabilities. Perhaps this ideal energy solution—unlimited pollution-free energy—will one day reward humanity. Meanwhile the carbon that we add to the atmosphere continues to produce irreversible climate change.

We may debate the advantages of the various alternative solutions but it is imperative that we act now to correct this dangerous situation. We have never paid the true cost of fossil fuel in the ongoing destruction of our precious climate, but with an alternative investment in development and support of solar and other renewable energy industries there can be a new economy based on existing, even free energy supplies.

Action

Life on Earth has evolved in harmony with the temperatures determined by the very unique atmospheric greenhouse effect, the physical laws of which have been understood in detail for over a hundred years. Now, we discover that certain of the greenhouse gas concentrations are being increased by the activities of our increased human population. The global warming that has been documented in average global temperature measurements on the Earth's surface is driving the numerous physical and biological climate changes that are beginning to be apparent.

Let us consider our possible responses. We may recognize the cause of extreme weather events and the need to decrease the emissions of greenhouse gases from burning fossil fuel; our government could enact a carbon tax. But this requires a significant change in government policy that is historically doubtful.

Or will we simply throw up our hands and declare 'excrement happens'. We will then have lost the opportunity to return to our normal climate. Nature will continue to follow the universal rules of the atmospheric greenhouse effect that we have described. There will be increased global warming and climate change and the environmental damage will be irreversible. We will also have decided to pay increasing billions for adaptation, destroying our present policies of economic growth.

In either response we must recognize that the natural laws of the environment will necessarily overrule the outdated human constructs of economics and government.

If we recognize that the natural atmospheric greenhouse effect continues to steadily respond to the increasing human-caused emissions of greenhouse gases, we may avoid the disastrous consequences of climate change. We may make an early choice to decrease our emissions of those greenhouse gases like CO_2 that control the thermostat on the atmospheric greenhouse effect. In fact, that is an entirely feasible possibility.

But that is not what I observe. The primary response is to aggressively pursue business as usual. This is aggravated by an economic situation; we fix it by subsidizing the methods that worked in the past. This requires more cheap energy. So those who are driven by early profit without regard to the next generation consciously disregard the science and accelerate drilling for oil and invest in future reservoirs of fossil fuels. Their lobbying efforts regarding government energy policy are driven by the need to protect those investments. Game over! That is a really stupid procedure in view of the need to cope with increasing costs of climate adaptation that are driven by the fossil fuel industries' failure to pay for their pollution. The fossil fuel industries have long operated for us to enjoy cheap power without concern for the consequences. No environmental costs were ever paid until some governmental regulations were required to respond to society's concern with health-related matters.

We must have government regulation of this blatant excess of corporate economics. It is time for an assessment of the costs on the carbon emissions related to electric power generation and vehicle transport that contribute to global warming and climate change. Those corporations argue that this will inevitably lead to increased costs to the consumers. The alternative is to face the increasing costs of adaptation to the human-caused climate change that will also ultimately endanger every aspect of society and life on this planet. We have a choice. We can face this dismal future of increasing costs of adaptation or invest these resources in the new industries of alternative energy. A new tomorrow including biofuels, nuclear, and clean inexhaustible solar energy is possible.

Recommendations

So what can a responsible citizen do? You've seen the list:

- change your light bulbs to CFLs

- install new double-or triple-paned windows

- paint your roof white

- upgrade insulation in your attic and crawlspace

- invest in photovoltaic panels for your rooftop or in a community or utility-sponsored solar garden

Some of these suggestions require a modest investment to conserve energy, but one can expect an eventual return in savings. However, this activity will only slow the global warming. We may expect rapidly increasing costs in adapting to disruptive climate change. Our leaders in environmental policy seem reluctant to move ahead when the citizens are apathetic and loath to follow. Let us identify and support leaders who will pursue a constructive human solution to this environmental dilemma. We must cease the domination of relatively unimportant political differences over attention to solutions to this dangerous climate change.

How do we do this? Be aware. Know what is happening in your neighborhood, your community, your state and your country. Educate yourself about climate change and what it will mean to you. It is easy to find a recommended reading list. Share your knowledge. Talk to family, friends, neighbors, and strangers. Use the social communications networks. When appropriate, join demonstrations like those organized by 350.org to insure visibility. Write letters to the elected officials in your community, county, state, and federal district. Finally, avoid the contrarian's trap of arguing the significance of yesterday's weather, or the temperature trend since the winter before last; these brief oddities do not constitute climate change. Concentrate on the controlling factors of atmospheric science.

It is naïve to expect the fossil fuel corporations to concede their responsibility for a major part of this environmental problem. With a citizenry ignorant or apathetic to the cause of climate change, these corporations are unlikely to independently volunteer to sacrifice their profits and invest heavily in alternative energy sources. And the present day practice of sound bites of unscientific opinions forestalls any intellectual public progress. On the other hand, if the natural laws of atmospheric science were to become common knowledge, our democratic society might hope to move toward constructive environmental solutions. It is the communication of this science that must be the initial step in controlling the climate change.

And you folks have the responsibility to learn about that science and to educate the next generation who will endure this new climate. Those who are leaders in education should ensure an exposure to the climate science from middle school years through college. A college graduate must understand the science of the global environment of humanity's future. A specific general education course requirement in all colleges is an obvious step to insure attention to our problem, but most of our university faculties either share the apathy of much of the public or enjoy substantial donations from donors allied with the fossil fuel industry. If we have truly evolved into a society with enough intelligence to recognize our influence on Nature, we have the responsibility to understand its fundamental laws and manage our lives in harmony with this unique planet.

We realize that our message will reach a very limited audience—like preaching to the choir. Will you heed the warning of things to come?

Cause "You ain't seen nothin' yet!" Nature will continue to follow the universal rules of the atmospheric greenhouse effect. There will be increased global warming and climate change. Adaptation will be increasingly costly and the environmental damage will be irreversible.

Conclusions

Our climate is going to follow Nature's laws involving the Atmospheric Greenhouse Effect. We have done a not-so-very-smart thing in knowingly increasing the greenhouse gas concentrations that have upset our planet's ability to maintain a stable climate.

In writing about our response to global warming and climate change, we have taken an initial attitude that there is a correction to be made to an undesirable situation. We must improve our care for the planet and the future of human civilization by controlling global warming and climate change. The problem has risen as the human population increased. It felt so righteous to 'go forth and multiply'! If we were back at a population of 4 billion, we could behave as irresponsibly as is our wont with negligible effect on Nature, short of an all-out nuclear war. But we are at 7 billion and growing. We can no longer behave as if we were not responsible for our actions.

There are those who question our right to insist that you waste your valuable time in studying a subject that you have never needed and to make drastic changes in your everyday life. But does Joe Blow really have the right to make our home an increasing target for wildfires, floods, or tornados so that he can continue to live in denial or ignorance?

Let us be perfectly clear. The choice isn't 'to believe or not to believe'. We have shown the human-caused increase of trace greenhouse gases like CO_2 functions as a control knob to produce global warming and subsequent climate change that is irreversible on our family's time scale.

Throughout history, scientists have found it useful to have a picture in their mind's eye to describe their observations. We hope it has been helpful for you to think of our atmosphere as composed of molecules and photons like microscopic ping pong balls that absorb and lose energy through collisions. Obviously we must use the language of mathematics to calculate the results and computers are necessary to do it quickly. We have presented the story of our planet's atmospheric greenhouse effect and its climate response. Now when you hear the predictions of computer models, you may have some better insight that these are describing reality. We have yet to see if humanity's choice for the story's ending will be happy or tragic.

Some individuals and organizations say the greenhouse effect is just a theory. We say, sure, a theory for the Earth's atmosphere that is supported by the very existence of human civilization on this planet! Newton's Law of Gravitation also works very well for Earth, and like Newton's theory of gravitation, the theory of heat radiation was developed more than a hundred years ago. Both are verified by observations. It is time to wake up and give attention to the established science of the greenhouse effect that now plays an increasing role in our everyday life.

The greenhouse effect is a law of nature, with unique beneficial circumstances for planet Earth. We can manage its effects by changing the concentration of greenhouse gases in our atmosphere. We are just beginning to learn this; some careful thought is required to avoid the painful consequences that may eventually become apparent. We believe you will agree that to ignore gravity and carelessly experience a fall off the roof is guaranteed to be sufficiently painful to be avoided. We have learned to conduct our affairs accordingly. We have to believe that you are intelligent enough that you understand our explanation that a certain combination of greenhouse gases has been critical for the climate of our ancestors and the plants and other creatures that supported them. It is our moral responsibility to preserve this climate for our descendants. So, get your rear in gear and manage accordingly!

Finally, may we remind you of the lessons of the ozone holes. Those spectacular destructive events were not predicted by the early atmospheric computer models because those model predictions were based on an incomplete understanding of the science. And now, present computer climate models do not include the emission rates of CO_2 and CH_4 from the melting permafrost or the release of methane clathrate from under the oceans; the science of the rate of this earth response to human caused warming is not yet well understood.

We face the uncertainty of arrival at these tipping points and the catastrophe of control by the centuries long lifetimes of CO_2, thus passing a point of no return to a livable climate. Our present apathy and procrastination move us ever closer to these potential disasters to human civilization.

Acknowledgements

We must first acknowledge the work of those scientists in the field—those glaciologists, oceanographers, meteorologists, and biologists—who have documented the existing climate changes in our backyards and in very remote areas of the world. Their reports have demonstrated the reality of climate change and have supported the analyses of laboratory scientists and the calculations and predictions of climate modelers. We are indebted to the statements and writings of renowned scientists like James Hansen and John Houghton who have made unequivocal declarations of the urgency of the danger of future climate changes. Numerous scientists from the laboratories of NOAA, NASA, and NCAR, as well as the science departments of universities with active programs in atmospheric science similar to those of Harvard, Princeton, Colorado, Arizona, and the laboratories of England, Germany, and New Zealand have contributed immensely to the vast published literature of the Earth's physical, chemical, and biological environment. Our modest contributions to research observations in atmospheric science and in communications to students and interested citizens have benefited greatly from the work and writings of those in the above scientific communities.

References

2014 National Climate Change Assessment. Highlights and full report available at http://nca2014.globalchange.gov/downloads

Alley, Richard. 2011. Earth: The Operator's Manual. W. W. Norton & Company.

Gore, Al. 2006. An Inconvenient Truth, The Planetary Emergency of Global Warming and What We Can Do About It. Rodale Books.

Gore, Al. 2009. Our Choice: How We Can Solve the Climate Crisis. Rodale Books.

Hamilton, Clive. 2010. Requiem for a Species: Why We Resist the Truth about Climate Change. Routledge.

Hansen, James. 2009. Storms of My Grandchildren. Bloomsbury USA.

Houghton, John. 2004. Global Warming, The Complete Briefing. Cambridge University Press.

Intergovernmental Panel on Climate Change (IPCC). 2007. Climate Change 2007: The Physical Science Basis.
Available at www.ipcc.ch/publications_and_data/publications_and_data_reports.htm

Mann, Michael E. and Lee R. Kump. 2008. Dire Predictions: Understanding Global Warming—The Illustrated Guide to the Findings of the IPCC. DK Publishing.

US National Academy of Sciences/The Royal Society (UK) joint report. 2014. Climate Change: Evidence and Causes. Available at www.nap.edu/catalog.php?record_id=18730

Walsh, Brian. 2012. Global Warming—The Causes—The Perils—The Solutions. Time Books.

Recommended Video

Al Gore. An Inconvenient Truth.

Al Gore. 24 hours of Reality

Richard Alley . Earth: The Operators Manual.

Arizona State University Origins Project. 2013
The Great Debate—Climate Change: Surviving the Future with Lawrence Krauss, James Hansen, Susan Solomon, Wallace Broecker, John Aston, and Sander Van Der Leeuw
Part 1: http://www.youtube.com/watch?v=XPaTAC29W2I
Part 2: http://www.youtube.com/watch?v=_pKZLHNYPbM

Lester R. Brown, Matt Damon. Journey to Planet Earth: Plan B

Planet Forward:PBS

Andrew Revkin, Jennifer Granholm, and Thomas Connnelly

350.org—Do the Math—The Movie

The Tipping Points. A documentary series produced by The Weather Channel. http://www.weather.com/tv/tvshows/tipping-points/main

Thin Ice: The Inside Story of Climate Science.
http://thiniceclimate.org/

Years of Living Dangerously. Showtime (available on DVD)

www.ingramcontent.com/pod-product-compliance
Lightning Source LLC
Chambersburg PA
CBHW071611170526
45166CB00003B/1053